Make: Bluetooth

WITHDRAWN

Bluetooth LE Projects for Arduino, Raspberry Pi, and Smartphones

Alasdair Allan, Don Coleman, Sandeep Mistry

MAKER MEDIA

SAN FRANCISCO, CA

Make: Bluetooth

by Alasdair Allan, Don Coleman, and Sandeep Mistry

Printed in the United States of America.

Published by Maker Media, Inc., 1160 Battery Street East, Suite 125, San Francisco, CA 94111.

Maker Media books may be purchased for educational, business, or sales promotional use. Online editions are also available for most titles (*http://safaribooksonline.com*). For more information, contact O'Reilly Media's institutional sales department: 800-998-9938 or *corporate@oreilly.com*.

Editor: Roger Stewart

Production Editor: Melanie Yarbrough

Copyeditor: Gillian McGarvey

Proofreader: Christina Edwards

Indexer: Last Look Editorial

Interior Designer: David Futato

Cover Designer: Jean Tashima

Illustrator: Rebecca Demarest

December 2015: First Edition

Revision History for the First Edition

2015-11-24:   First Release

See *http://oreilly.com/catalog/errata.csp?isbn=9781457187094* for release details.

978-1-457-18709-4

[LSI]

Table of Contents

Preface

One of the drivers behind the recent explosive growth in the Internet of Things has been the Bluetooth Low Energy (LE) standard. What makes it so appealing is the ubiquity of smartphones—both Apple and otherwise—with support for the standard, which means that thing-makers no longer have to worry about a display or user interface. And that means that things like smart lightbulbs can look a lot more like lightbulbs rather than a computer that happens to have a light attached to it.

Bluetooth LE is very different from classic Bluetooth—in fact, pretty much the only thing that is the same is the name. You're probably used to thinking about radios as sort of like serial connections working similarly to a phone call between two phones—once you establish a connection, each person talks as the other listens and vice versa. They stay connected, even if neither is saying anything, until one hangs up and the call is ended.

In systems like these, data is transferred using a queue, and when data is read by the receiver, it's erased from the queue—just as once my words reach your ears over the phone, they're out of the communications channel. Effectively, this is how classic Bluetooth works.

What Makes Bluetooth LE Unique?

Instead of communicating via a point-to-point connection like a phone, a *Bluetooth LE radio* acts like a community bulletin board, with each radio acting as either a board or a reader of the board.

If your radio is a bulletin board—called a *peripheral device* in Bluetooth LE parlance—it posts data on its board for everyone in the community to read. If your radio is a reader—called a *central device* in Bluetooth LE terms—it can read from any of the boards (the peripheral devices) that have information it cares about.

If you don't like that analogy, you can also think of peripheral devices as the servers in a client-server transaction. Similarly, central devices are the clients of the Bluetooth LE world because they read information from the peripherals.

But I Like Serial Connections!

Most (perhaps all?) of the Bluetooth LE radios breakout boards available to makers right now—the RedBearLab BLE mini and the Adafruit Bluefruit LE, for instance—pretend to look like serial devices for simplicity's sake and present a UART service to the user. Effectively these radios are "faking" old-style serial communication on top of the underlying bulletin board paradigm. It's a hack, and not a bad one.

Serial over Bluetooth LE makes the transition from classic Bluetooth to Bluetooth LE easier for people who are used to Serial Port Profile (SPP) and UART. However, there's a downside. It doesn't take advantage of Bluetooth LE. These Bluetooth serial services are just stuffing data across generic transmit and receive pipes. If a device uses serial, it also needs to define a protocol for the data that is sent and received.

Bluetooth LE offers device makers the ability to create Bluetooth LE devices with self-describing services. If the characteristics are well designed and the descriptors make sense, you can use services without documentation. (An example of this is using a Smartbotics Lightbulb with the LightBlue iOS app.)

Contrast this with other devices using SPP, like services where you need to learn the details of which bytes to send over the wire to turn on an LED and change its color. These devices need really good documentation and/or additional libraries to do simple stuff like set the color of a LED.

Returning to our bulletin board example, we're creating a board (the *service*) that has a sticky note attached (known as a *characteristic* in Bluetooth LE parlance), which we can read, letting us know if the LED is on or off, or write to—allowing us to control the LED.

Building a Custom Service

Unfortunately, until recently, building custom services for Bluetooth LE has been fairly complicated and not for the faint-hearted. However, it's getting simpler as several good tools now exist to do most of the heavy lifting for you.

In light of that, we decided to look at one platform—the Nordic Semiconductor radios—and figure out a complete toolchain that would allow you to build a custom service for those radios, and make use of that service from an Arduino project. We picked this particular radio because it's readily available and there is good library support.

This book is the result.

Who Should Read This Book?

This book provides an introduction to the topic of how to build and deploy Bluetooth LE sensors and devices. It takes a hands-on approach and teaches you how to build things and how to use them, not just the protocols and architecture behind the standard.

If you're a programmer or a maker who wants to get started with Bluetooth LE, this book is for you.

What You Should Already Know

This book is intended as an introduction to working with Bluetooth LE, and it assumes a technical background. However, we walk you through installing and using all the software and development environments you'll need, including how to get started with the Arduino, Raspberry Pi, Node.js, and PhoneGap.

What You Will Learn

This book will guide you through building a series of connected projects—from lightbulbs, to locks, to beacons, and drones. It will walk you through your first hardware prototypes, show you how to improve them, and teach you how to build Bluetooth LE connected devices for the Internet of Things.

What's In This Book

Chapter 1, *Introduction*
> This chapter talks about the Bluetooth LE standard and the concepts you'll need to know about before you can start building Bluetooth LE devices.

Chapter 2, *Getting Started*
> This chapter walks you through setting up the tools and software you'll need to set up your development environment.

Chapter 3, *Smart Light Switch*
> This chapter shows you how to build a smart light switch that not only lets you turn the light on or off using the switch but also remotely via Bluetooth LE. The switch knows its current status—in other words, whether the bulb is on or off—and will send out a notification if that status changes.

Chapter 4, *BLE Lock*
> This chapter shows you how to build a lock that can be opened using your phone via Bluetooth LE. We will also walk through writing a mobile app that will run on iOS or Android using PhoneGap, which will control the lock.

Chapter 5, Bleno Lock

> This chapter re-creates the Bluetooth LE lock built in the previous chapter but this time using Node.js on the Raspberry Pi. It uses the same lock service as the original Bluetooth LE lock so it can be controlled using the same mobile app.

Chapter 6, Weather Station

> This chapter walks you through building a Bluetooth-LE-enabled weather station that can measure temperature, humidity, and pressure.

Chapter 7, NeoPixel Lamp

> This chapter will show you how to build an RGB lamp that is controllable from your phone using an Arduino and a NeoPixel Ring with 16 LEDs.

Chapter 8, SensorTag Remote

> This chapter shows you how to turn a TI SensorTag into a remote control for your computer, triggering an action on your computer when one of the SensorTag buttons is pressed.

Chapter 9, HID over GATT

> This chapter explores the Human Interface Device (HID) profile and shows you how to build a Bluetooth LE volume control.

Chapter 10, Beacons

> In this chapter we explore Bluetooth LE beacons, showing how to create them using Node.js and detect them using smartphones.

Chapter 11, Drones

> This chapter shows you how to control a Parrot Rolling Spider drone over Bluetooth LE using a computer and Node.js.

Chapter 12, Going Further

> This chapter provides a collection of pointers to more advanced material on the topics we covered in the book, and material covering some of those topics that we didn't manage to talk about in this book.

Conventions Used in This Book

The following typographical conventions are used in this book:

Italic

> Indicates new terms, URLs, email addresses, filenames, and file extensions.

`Constant width`

> Used for program listings, as well as within paragraphs to refer to program elements such as variable or function names, databases, data types, environment variables, statements, and keywords.

Constant width bold

Shows commands or other text that should be typed literally by the user.

 This element signifies a general note, tip, or suggestion.

 This element indicates a warning or caution.

Using Code Examples

This book is here to help you get your job done. In general, if example code is offered with this book, you may use it in your programs and documentation. You do not need to contact us for permission unless you're reproducing a significant portion of the code. For example, writing a program that uses several chunks of code from this book does not require permission. Selling or distributing a CD-ROM of examples from Make: books does require permission. Answering a question by citing this book and quoting example code does not require permission. Incorporating a significant amount of example code from this book into your product's documentation does require permission. All the code from this book is available on GitHub at *https://github.com/MakeBluetooth*.

Support material for the book is also available at the book's website (*http://makeblue tooth.com*).

We appreciate, but do not require, attribution. An attribution usually includes the title, author, publisher, and ISBN. For example: "*Make: Bluetooth* by Alasdair Allan, Don Coleman, Sandeep Mistry (O'Reilly). Copyright 2016 Alasdair Allan, Don Coleman, Sandeep Mistry, 978-1-457-18709-4."

If you feel your use of code examples falls outside fair use or the permission given above, feel free to contact us at *bookpermissions@makermedia.com*.

Safari® Books Online

 *Safari Books Online is an on-demand digital library that delivers expert content in both book and video form from the world's leading authors in technology and business.*

Technology professionals, software developers, web designers, and business and creative professionals use Safari Books Online as their primary resource for research, problem solving, learning, and certification training.

Safari Books Online offers a range of plans and pricing for enterprise, government, education, and individuals.

Members have access to thousands of books, training videos, and prepublication manuscripts in one fully searchable database from publishers like O'Reilly Media, Prentice Hall Professional, Addison-Wesley Professional, Microsoft Press, Sams, Que, Peachpit Press, Focal Press, Cisco Press, John Wiley & Sons, Syngress, Morgan Kaufmann, IBM Redbooks, Packt, Adobe Press, FT Press, Apress, Manning, New Riders, McGraw-Hill, Jones & Bartlett, Course Technology, and hundreds more. For more information about Safari Books Online, please visit us online.

How to Contact Us

Please address comments and questions concerning this book to the publisher:

Make:
1160 Battery Street East, Suite 125
San Francisco, CA 94111
877-306-6253 (in the United States or Canada)
707-639-1355 (international or local)

We have a web page for this book, where we list errata, examples, and any additional information. You can access this page at *http://bit.ly/make-BT*.

Make: unites, inspires, informs, and entertains a growing community of resourceful people who undertake amazing projects in their backyards, basements, and garages. Make: celebrates your right to tweak, hack, and bend any technology to your will. The Make: audience continues to be a growing culture and community that believes in bettering ourselves, our environment, our educational system—our entire world. This is much more than an audience, it's a worldwide movement that Make is leading we call it the Maker Movement.

For more information about Make:, visit us online:

- Make: magazine (*http://makezine.com/magazine*)
- Maker Faire (*http://makerfaire.com*)
- Makezine.com (*http://makezine.com*)
- Maker Shed (*http://makershed.com*)

To comment or ask technical questions about this book, send email to *bookquestions@oreilly.com*.

Acknowledgements by Alasdair Allan

Everyone has one book in them, but this isn't mine. Depending how you count them, this is my ninth book. But every book is different, and they do not write themselves. So I'd like to thank my co-authors Don Coleman and Sandeep Mistry, and my editors at Make:, Brian Jepson and Roger Stewart, for holding my hand throughout the process.

I very much want to thank my wife, Gemma Hobson, for her continued support and encouragement. Those small, and sometimes larger, sacrifices an author's spouse routinely has to make don't get any less inconvenient the second, or third, or the n'th time around. I'm not sure why she lets me write—perhaps because I claim to enjoy it so much. Thank you, Gemma. Finally to my son Alex who, seven years on from my first book, is actually almost old enough to read this one.

Acknowledgements by Don Coleman

I'd like to thank my wife, Meghan, and son, Liam, for their support and patience during the many hours I spent working on this book. Tom Igoe was very helpful by working through ideas and testing early versions of code and libraries with his students. Guan Yang's work on Arduino nRF8001 provided a huge kickstart in getting custom BLE services running on Arduino without proprietary tools. Brian Jepson, Roger Stewart, and the team at Make: deserve a shout-out for their patience with us and their persistence in getting this book published. I appreciate the support of my colleagues at Chariot Solutions, LLC. Lastly, it's been awesome working with my talented co-authors Alasdair and Sandeep.

Acknowledgements by Sandeep Mistry

This is the first book I've been involved in. I would like to thank my co-authors Alasdair Allan and Don Coleman, and the editors Brian Jepson and Roger Stewart for guiding me through the process and providing valuable feedback throughout its production.

Introduction 1

In Chapter 2, we'll install the tools we need to write code and deploy services. But before we do that, we need to talk about Bluetooth LE, and make sure your machine can talk to it, too.

This chapter talks about some of the jargon you'll need to understand to work with Bluetooth LE. The following chapter walks you through installing all of the software and configuring the hardware you'll need for the projects in this book. After working through these two chapters, you should probably proceed with Chapter 3, which is a solid introduction to working with Bluetooth and the Arduino. However, after that you should be able to pick and choose among the rest of the projects in the book.

Talking About Bluetooth LE

Bluetooth LE divides the world into peripheral and central devices. *Peripheral devices* are things like sensors; they're typically small, low-powered, and resource-constrained. *Central devices* are things like mobile phones and laptops, although these can usually also operate in peripheral mode.

Peripherals can operate in two modes: either by broadcasting or being directly connected to a central device. The broadcast mechanism is one of the big differences between Bluetooth LE and classic Bluetooth; it allows data to be sent out by the peripheral to any device in range.

The real-world range of a Bluetooth LE device depends on the transmitting power of the radio. Since higher transmitting power means more battery is required, Bluetooth LE is, unsurprisingly, a short-range standard. While it's perfectly possible to have a real-world range of greater than 30m (about 100ft), a more typical operating range is between 2 and 5m (around 5 to 15ft).

This means that a Bluetooth LE peripheral device doesn't necessarily need to be "paired" with a central device in order to transfer data. In Bluetooth LE, we speak of it as "connected" rather than paired as we did with Bluetooth 2.1. In broadcast mode, the peripheral will periodically send out advertising packets, available to any device that's looking for them, for devices acting as "observers."

The standard *advertising packet* describes the broadcasting device and its capabilities but is also capable of including custom information—sensor data, for instance—that you want to broadcast.

Broadcasting data from your peripheral is a good choice if you're building something like a weather station where the data isn't sensitive. However, there is no provision for security when broadcasting, so for personal data, the central device should connect to the peripheral, not the other way around.

Connections are exclusive. This means that a peripheral cannot be connected to more than one central device at a time. When a central device connects to a peripheral, the peripheral will stop advertising itself, and other devices will not be able to see it or connect to it until the first connection is terminated. However, whereas a peripheral can only be connected to one central device, a central device can be connected to more than peripheral at the same time.

The Bluetooth 4.1 specification removed the restriction that peripheral devices can only be connected to a single central device. Going forward, a peripheral can be connected to multiple central devices simultaneously. However, there are still many devices and chipsets that remain limited in this fashion.

If you need to exchange data between the peripheral and the central device, then you need to establish a connection between the two devices.

Protocols and Profiles

On top of the protocols that make up the Bluetooth LE standard, the specification defines what are called *profiles*. These are either the basic modes of operation needed by all Bluetooth LE devices, like the Generic Access Profile and Generic Attribute Profile, or profiles covering specific use cases (*https://developer.bluetooth.org/TechnologyOverview/Pages/Profiles.aspx*) such as the Heart Rate Profile.

The GAP

The *Generic Access Profile* (GAP) defines roles for devices, including the peripheral and central roles we mentioned in the last section, alongside advertising and discovery.

There are two ways to advertise data using GAP: advertising data and scan response packets. While both packets use the same payload format, and consist of up to 31 bytes of data, only the advertising data packet is mandatory. It is sent out at a preset advertising interval —the longer the interval, the less power used—on receipt listening devices can request the scan response packet with additional data if it exists.

Using custom advertisement data in the broadcast packets is how both the iBeacon and Eddystone standards are implemented; see Chapter 10 for more information.

Once a connection with the peripheral has been made, you will use GATT services and characteristics to communicate with the peripheral device, and advertising will stop until the connection is terminated.

The GATT

The *Generic Attribute Profile* (GATT) defines how Bluetooth LE transfers data back and forth between peripheral and central devices. It defines profiles, which are collections of services. Each service has characteristics, which contain data.

Roles change when moving from GAP to GATT. GATT defines two roles: client and server.

It may seem backwards, but peripheral devices are known as *GATT servers* and the more powerful central devices are *GATT clients*. Think of it this way: the server has data and the client wants data. All connections between the devices are initiated by the client.

After connecting, the client can get a list of services offered by the server. Before connecting, the central device had a potentially incomplete list of services from the advertising data.

Services and Characteristics

Services are used to break up the data into logically associated chunks, and consist of a collection of characteristics. *Characteristics* are the containers that hold the data associated with a service. Both services and characteristics are identified by a unique identifier, known as an UUID; see "UUIDs".

Characteristics contain at least two attributes: a *characteristic declaration*, which contains metadata about the data, and the *characteristic value*, which contains the data itself. The characteristic can also contain additional descriptors to expand on the meta data. Together, the declaration, value, and any optional descriptors form a bundle than make up a characteristic.

Characteristics can be defined as read or write. Characteristics are read by the client using a read request, with the returned value of the request being the characteristic value. Characteristic values can be written using a *write request*. The server returns a confirmation after the value is written. There is an additional write property called the *write command*. When a characteristic value is written with a write command, the server does not send any response back to the client. The write command is sometimes called write without response.

Two additional properties are *notify* and *indicate*. Both of these are server-initiated communication. A client subscribes to be notified when a characteristic's value changes. When a change occurs, the server notifies the client by sending the new value. An indication is similar to a notification, except that the client must acknowledge the receipt of the indication.

Characteristics can have multiple properties. For example, one characteristic could allow `read`, `write`, and `notify`.

UUIDs

Bluetooth uses Universally Unique Identifiers (UUIDs) for many things, including services and characteristics. Bluetooth services that have been approved by the Bluetooth Special Interest Group (*http://bit.ly/1giCbMM*) are assigned 16-bit UUIDs. All other services and characteristics must use 128-bit UUIDs, which can be generated with tools such as `uuidgen`, as shown here:

```
$ uuidgen
437121E5-A6F0-43F9-8F8F-4AB73D6CC3EB
```

In this book, we use 16-bit UUIDs. Technically we're breaking the rules, but for us it's a lot easier to type `DCF8` than `391CDCF8-4BD4-4507-BE23-B57DFD1F870B`. See Table 1-1 for a list of officially approved 16-bit UUID ranges.

Table 1-1   *Table 1-1. Officially approved 16-bit UUID ranges*

UUID	Description
00xx	Namespace Descriptors
18xx	Services
27xx	Units

UUID	Description
28xx	Declarations
29xx	Descriptors
2Axx	Characteristics

For the LED examples, we reuse SmartBotics' service for their RoboSmart lightbulb (*http://smartbotics.com*). For the button examples, we reused Texas Instruments' Simple Key Service (*http://bit.ly/simple-key-service*) from their Sensor Tag. For the thermometer examples, we create some 16-bit UUIDs.

It's fine to reuse UUIDs if existing services and characteristics meet your needs. If you're making your own services, use 128-bit UUIDs. See the Bluetooth Developer Site (*https://developer.bluetooth.org*) for more information.

An Example Service

Pulling this together, let's look at a typical example service for a lightbulb, as shown in Table 1-2 and illustrated in Figure 1-1.

This is part of the service definition advertised by a RoboSmart Light Bulb (*http://amzn.to/20ckMeO*). As you can see, despite not being an officially approved service definition, like many other manufacturers, RoboSmart uses 16-bit UUIDs.

Table 1-2 *Lightbulb Service (FF10)*

Characteristic	UUID	Property	Value	Comments
Light switch	FF11	Read, Write	1	1 on, 0 off
Dimmer setting	FF12	Read, Write	0×7F	0×00 to 0×FF
Power consumption	FF16	Read	340	Watt Hours

This is part of the service definition advertised by a RoboSmart Light Bulb (see Figure 1-1).

Figure 1-1 *Exploring the Lightbulb service*

Making Sure Your Machine Has Bluetooth LE

Support for Bluetooth LE is available on most major OS platforms. However, if your laptop or desktop hardware doesn't support Bluetooth LE, there is a large number of Bluetooth LE USB dongles available. Many are based around the same Broadcom BCM20702A0 chipset; for this book, we used the IOGEAR GBU521 Bluetooth 4.0 USB Micro Adaptor (*http://amzn.to/1Jdbetl*), which can be picked up for about $10 to $15 on Amazon and other on-line retailers. USB adapters with the CSR8510 chipset also work well.

OS X

All Apple laptop and desktop machines built after 2011 have Bluetooth LE hardware built-in, though if you're using a Mac built before that and running Mountain Lion or later, you're probably still okay. You can pick up a Bluetooth LE USB adaptor and it should work out of the box.

If you're running a pre-Mountain-Lion distribution of OS X, it's still possible to get most Broadcom-based Bluetooth LE adaptors to work, but it's probably going to be a lot harder. For Lion, you'll have to go into the /System/Library/Extensions/ folder and make changes inside the IOBluetoothFamily.kext, adding the current Product ID and Vendor ID values of your USB dongle before reloading the kernel extension. For those running Pre-Lion distributions of OS X, things are even more difficult.

Apple iOS

Support for Bluetooth LE on the iPhone and iPad has existed since iOS 5, although we'd recommend iOS 7 as the minimum version because it introduced iBeacon support to the operating system. Hardware support for Bluetooth LE was introduced with the iPhone 4s, iPad (third generation), and the iPod touch (fifth generation). All current Apple iOS devices support the standard.

Linux

Linux uses the BlueZ service to talk to Bluetooth devices, and Bluetooth LE has been supported since version 4.93.

Linux users should take a look at the instructions in "Setting Up Raspberry Pi" where we talk about installing the BlueZ service to work with the Bluetooth 4.0 USB adaptors.

Android

Support for Bluetooth LE in the core Android OS was introduced in Android 4.3, although version 4.4 introduced numerous bug fixes and is probably the minimum recommeded version. Before this, though some manufacturers shipped hardware that was Bluetooth-LE-compatible, there was no software support, which caused many manufacturers to introduce their own libraries to support it—which were unfortunately all incompatible.

Microsoft Windows

Microsoft Windows 8 and above has built-in support for Bluetooth LE. Unfortunately, Windows XP, Vista, and Windows 7 only support Bluetooth 2.1, and while it's theoretically possible to get these earlier versions to talk to Bluetooth LE devices—at least for Windows 7, it's an extremely challenging problem that's well beyond the scope of this book. If you're working with Bluetooth LE under Windows, you should at least be running Windows 8.

What Haven't We Told You About Bluetooth LE?

Lots. The Bluetooth LE specification is a sprawling mess of interlocking documents (*https://www.bluetooth.org/en-us/specification/adopted-specifications*) that runs to thousands of pages; the core standards document (*https://www.bluetooth.org/DocMan/handlers/Down loadDoc.ashx?doc_id=286439*) is over 2,700 pages on its own. What we've given you here is the barest outline, a sketch, of how the standard works.

However, it's enough that you should now be able to go and confidently build projects using Bluetooth LE to actually do things in the real world. You can pick up the architecture and theory behind the protocols as you go along. But you have enough to get started.

Getting Started | 2

Before you start writing code, you'll need to do some housekeeping. First, you'll need to install Node.js, the Arduino IDE, and libraries to support the Bluetooth LE chipset—the nRF8001 from Nordic Semiconductor—that we'll be using in the products throughout the book.

Let's get these housekeeping tasks out of the way so you can get to the interesting bit—the code—as soon as possible.

The Arduino

The Arduino started off as a project to give artists access to embedded microprocessors for interaction design projects, but it may well end up in a museum as one of the building blocks of the modern world. It allows rapid, cheap prototyping for embedded systems. It turns what used to be fairly tough hardware problems into simpler software problems (and we know that once problems are in the realm of software, they become almost exponentially easier).

The Arduino—and the open hardware movement that has grown up with it and to a certain extent around it—is enabling a generation of high-tech tinkerers both to break the seals on proprietary technology and to prototype new ideas with fairly minimal hardware knowledge. This Maker renaissance has led to an interesting growth in innovation. People aren't just having ideas; they're doing something with them.

The Board

The revision of the Arduino board we're using in this book is known as the Arduino Uno, which is shown in Figure 2-1. It is based on the ATmega328 microcontroller, and it, along with its many clones and compatibles, is one of the most widely available and easily found microcontroller boards on the market today.

Figure 2-1 *The Arduino Uno board with the ATmega328 microcontroller*

It has 14 digital input/output pins, six of which can be used as PWM outputs, along with six more analog input pins. Table 2-1 shows the technical specifications of the Arduino Uno.

Table 2-1 *Technical specifications of the Arduino Uno board*

Arduino Uno	
Microcontroller	ATmega328
Operating Voltage	5 V
Input Voltage (recommended)	7 – 12 V
Input Voltage (limits)	6 – 20 V
Digital I/O Pins	14 (6 provide PWM)
Analog Input Pins	6
DC Current per I/O Pin	40 mA
DC Current for 3.3V Pin	50 mA
Flash Memory	32 KB
SRAM	2 KB
EEPROM	1 KB
Clock Speed	16 MHz

The Arduino platform has everything needed to support the on-board microcontroller. It is attached directly to your computer for programming using a USB connection, and it can be powered via the same USB connection or with an external power supply if you want to detach the board from your computer after you've programmed it.

Powering the Board

The Arduino Uno can be powered via the USB connection, or with an external power supply. Unlike previous generations of the Arduino, the power source is selected automatically. If you're using an earlier model, you will have to manually change between USB and external power sources using a jumper on the board itself; this jumper is usually located between the USB and power jacks.

The board can operate on an external supply of 6 to 20 volts. However, if supplied with less than 7V, the 5V pin may supply less than nominal voltage and the board may become unstable. If using more than 12V, the voltage regulator may overheat and damage the board. The recommended range is therefore between 7 and 12 volts.

Input and Output

Each of the 14 pins on the Uno can be used as an input or output. They operate at 5V, with each pin having an internal pull-up resistor (disconnected by default) of 20 to 50 kOhms. The maximum current a pin can provide is 40mA.

 Some pins have specialized functions. Perhaps the most important of these for the purposes of this book are pins 0 and 1. These pins can be used to receive (RX) and transmit (TX) TTL serial data. These pins are connected to the corresponding pins of the ATmega8U2 USB-to-TTL serial chip.

Communicating with the Board

The ATmega328 provides UART TTL serial communication at 5V, which is available on digital pins 0 (RX) and 1 (TX). The Ardunio Uno has an ATmega8U2 chip on-board that redirects this serial communication over USB, allowing the Arduino to appear as a virtual serial port to software on your laptop.

Installing the Arduino IDE

We will use the Arduino IDE to develop code, as shown in Figure 2-2.

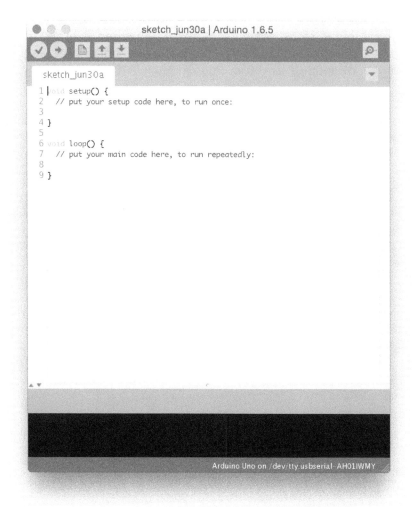

Figure 2-2 *The Arduino IDE*

 The latest version of the Arduino development environment (http://ardui no.cc/en/Main/Software) is available. At the time of writing, this was Arduino 1.6.5.

Go ahead and download the latest version of the development environment from the Arduino website (*http://arduino.cc*).

Installing on OS X

Like most Mac applications, the development environment comes as a disk image (.dmg file) that should mount automatically after you download it. If it doesn't, double-click on it to open it manually. After it is open, just click-and-drag the `Arduino.app` application into your `/Applications` folder. Double-click on the application to open it.

If you're using a Mac when you connect the board to your computer, depending on the version of OS X you're running, a dialog box may appear telling you that a new network interface has been detected. Just hit the Apply button. Even though the new interface will claim to be "Not Configured," if you inspect it in System Preferences, you will see that it is working correctly.

Installing on Linux

The Arduino development environment has been packaged and can be installed by using the package manager. On Debian, type:

```
$ sudo apt-get install arduino
```

Detailed instructions for other flavors of Linux are available on the Arduino Playground (*http://playground.arduino.cc/Learning/Linux*).

For the Aruino Uno, you need `librxtx-java version 2.2pre2-3` *or later installed.*

Installing on MS Windows

The development environment comes as a ZIP file. Download it, unzip it, and double-click on it to open the folder, and then double-click on the application to open it.

If you're running Windows when you connect the board to your computer, driver installation should begin automatically. However, if you're running Windows 7, Vista, or XP, it may fail. If so, click on the Start Menu and open the Control Panel. Navigate to System and Security, and click on System. Once the System window is open, go ahead and open the Device Manager. Under Ports (COM & LPT), or possibly under Unknown Devices, you should see an entry named Arduino UNO (COMx). Click on the entry and choose Update Driver Software and Browse My Computer for Driver Software. Navigate to the folder containing the development environment and choose the Arduino Uno driver named `Ardunio UNO.inf`.

Connecting to the Board

Connect the Arduino to your computer with an appropriate USB cable. In the case of an Arduino Uno, you'll need a USB-A (computer) to USB-B (Arduino) cable, the same sort needed for most USB printers. The green power LED (labeled PWR) should turn on.

After connecting the board to your computer, go to the Tools → Board menu item in the Arduino development environment (see Figure 2-3) and select your board from the list in the drop-down menu.

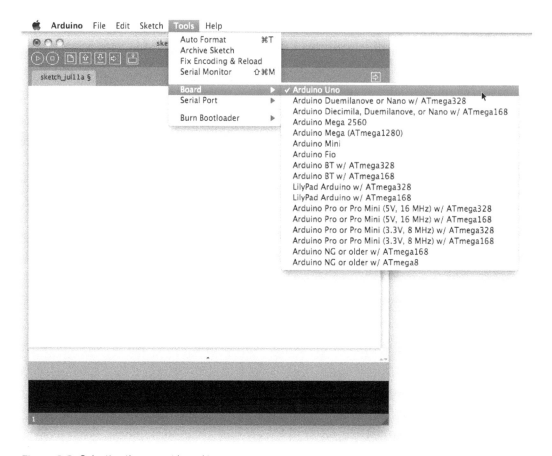

Figure 2-3 *Selecting the correct board type*

Then go to Tools→Serial Port and select the correct serial port for your board, as shown in Figure 2-4. On the Mac, the name will start with /dev/tty.usbmodem, whereas on Windows it will show up as a COM port.

Figure 2-4 *Choosing the correct serial port*

If you're unsure which serial port corresponds to your board, you can always unplug and then re-plug-in the USB cable connecting your computer to the Arduino board to see how the menu changes.

Installing the BLE Peripheral Library

Now that we have the Arduino IDE installed, we need to install the Bluetooth LE Peripheral Library. This is an Arduino library for creating custom BLE peripherals with Nordic Semiconductor's nRF8001 or nR51822.

The installation of this library has been simplified in the 1.6.x release of the Arduino IDE. Open the Arduino IDE and choose Sketch → Include Library → Manage Libraries… from the menu bar to open the Library Manager, as shown in Figure 2-5.

Figure 2-5 *The Arduino Library Manager window*

Go ahead and type **BLEPeripheral** into the search box on the top-left corner of the window. This will bring up the BLEPeripheral library. Select the library in the main pane and click on the Install button. The latest version of the library should be automatically installed and made available for use in your code.

 Throughout the book, we'll use the Adafruit Bluefruit LE board based around the Nordic Semiconductor nRF8001 chipset. As we mentioned in the Preface, the nRF8001 Arduino library (https://github.com/adafruit/ Adafruit_nRF8001) from Adafruit only supports a UART service. By using the BLEPeripheral library, we can use the same hardware but in a much more flexible and powerful way.

Setting Up Raspberry Pi

Some of the examples in this book use the Raspberry Pi. While the Pi doesn't have Bluetooth LE built in, there are many USB Bluetooth LE adaptors available. We use the IOGEAR GBU521 adaptor, and highly recommend it.

Installing BlueZ

Linux uses the BlueZ package to provide support for the core Bluetooth layers and protocols.

The stock BlueZ that's available via apt-get on Raspbian Wheezy is version 4.99, which can be installed by running:

```
sudo apt-get install bluez
```

 BlueZ 5.x is not yet considered stable, so for now you should continue to use BlueZ 4.x

Verifying the Bluetooth LE

Make sure your Bluetooth LE USB adaptor is plugged into the Pi, and then type:

```
$ hciconfig
```

You should see something like this:

```
hci0:   Type: BR/EDR  Bus: USB
    BD Address: 00:1A:7D:DA:71:0C  ACL MTU: 310:10  SCO MTU: 64:8
    UP RUNNING PSCAN
    RX bytes:979 acl:0 sco:0 events:43 errors:0
    TX bytes:910 acl:0 sco:0 commands:43 errors:0
```

Just to check that things are working correctly, type:

```
$ sudo hcitool lescan
```

Then you should see any Bluetooth LE peripherals that are within range, such as:

```
LE Scan ...
78:C5:E5:6C:D5:EA (unknown)
78:C5:E5:6C:D5:EA Hone
```

Hit ^C to stop the scan.

Node.js

Node.js is a system built on top of the V8 Javascript Engine. It uses an event-driven, non-blocking I/O model and has an extensive package management system called npm. It's open source and uses a cross-platform runtime environment for server-side and networking applications. Node.js applications are written in JavaScript, and are run within the Node.js runtime environment.

We're going to install the Node.js engine, and then install the relevant dependencies for the Bluetooth LE libraries we'll use throughout the book.

Installing Node.js

Probably the easiest way to install Node.js for those running OS X or MS Windows is to download the installer (*http://nodejs.org/#download*) directly from the Node.js website.

On Linux and Raspberry Pi

Node.js is available from the NodeSource Debian (*https://nodesource.com/*) and Ubuntu binary distributions repository. On a Debian/Ubuntu distribution you can install Node.js as follows:

```
$ curl --silent --location https://deb.nodesource.com/setup_0.12 | sudo bash -
$ sudo apt-get install --yes nodejs
```

Optionally, you can install build tools that allow you to compile and install native add-ons from npm, like this:

```
$ sudo apt-get install --yes build-essential
```

Installing Libraries with npm

Node.js libraries can be installed using the npm package manager that is installed alongside Node.js. You can install packages from the command line as shown here:

```
$ npm install PACKAGE_NAME
```

where PACKAGE_NAME is the name of the package you want to install. Some packages with command-line tools, like PhoneGap, can be installed globally by adding the -g flag. This is the exception. Installing packages locally is preferred. A full list of the packages available via npm can be found at npmjs.org (*https://www.npmjs.com/*).

 OS X users might be prompted to install the XCode or the command-line developer tools. This is only necessary if you will install Node.js packages that use native bindings and need to be built from source, such as noble *and* bleno.

Setting Up Dependencies for noble and bleno

OS X

Install Xcode from the Mac Appstore (*http://apple.co/1MXyWXf*). After it is installed, make sure to open it once to accept the license agreement.

Raspberry Pi and Linux

Both noble and bleno require some dependencies for the native Node.js modules on which they depend. They can be installed using the following command:

```
sudo apt-get install bluetooth bluez-utils libbluetooth-dev libudev-dev
```

Installing PhoneGap

PhoneGap allows us to build cross-platform mobile applications with HTML, JavaScript, and CSS. These applications are packaged and delivered to phones as native apps through the Apple AppStore or Google Play.

At its simplest, PhoneGap looks like a web browser embedded in a native app, but it's really more than that. PhoneGap provides JavaScript APIs to access native functions like Camera, Contact, and Accelerometer. These JavaScript APIs work across platforms. You can run the same code on iOS, Android, and Windows Phone. In addition to the built-in APIs, PhoneGap also provides a framework that allows anyone to add new APIs.

PhoneGap development requires command-line tools to be installed on your machine. It requires Node.js plus the native SDKs for each platform to be installed. It requires a Mac if you want to develop for iOS. iOS development also requires that you join the iOS Developer Program.

There are many barriers to entry here: tool setup, platform restrictions, developer programs. These are all things that can be easily conquered if you're going to develop apps, but it is a bit much to include in a book that's supposed to teach you about Bluetooth LE. Instead, we're going to try a different approach.

PhoneGap Developer App

The PhoneGap team has released the PhoneGap Developer App (*http://app.phonegap.com/*). The idea is that you can install Node.js and PhoneGap on your computer, and you can develop apps without installing native SDKs or worrying about signing code.

The PhoneGap Developer App pairs with your computer and loads your program via WiFi. You can't distribute these apps, but it works well for development. It is especially good for developing an iOS application on a Windows or Linux computer, which is something that's otherwise impossible to do without jailbreaking.

First, you need to install the PhoneGap command-line interface. This command-line tool is installed using npm:

```
$ sudo npm install -g phonegap
```

Then download the PhoneGap Developer App from the iOS App Store (*http://apple.co/1MXz49f*) or Google Play (*http://bit.ly/1MXz4WN*) onto your phone.

Bluetooth Low Energy Plugin

Bluetooth Low Energy support for PhoneGap is provided by a plugin (*https://github.com/don/cordova-plugin-ble-central*). The examples in the following chapters show you how to use the plugin. The GitHub project has API documentation (*http://bit.ly/ble-api*) and additional code examples (*http://bit.ly/ble-examples*).

PhoneGap Versus Cordova

What's the difference between PhoneGap and Cordova? PhoneGap was originally developed by Nitobi, which was later bought by Adobe. Before Nitobi was purchased by Adobe, they donated the PhoneGap codebase to the Apache Software Foundation to ensure it would remain open source. As part of the transition from Nitobi to Apache, the project was renamed as Cordova. PhoneGap is now Adobe's distribution of Cordova.

PhoneGap is powered by Apache Cordova. Adobe continues to add additional tools like the PhoneGap Developer App (*http://app.phonegap.com*) and PhoneGap Build (*https://build.phonegap.com*). We're using PhoneGap here because it allows us to use the PhoneGap Developer App.

Platform Tools

When working with beacons in Chapter 10, you'll need the Cordova iBeacon plugin. Since this plugin is not included in the PhoneGap Developer App, you'll need to run the full version of PhoneGap. The full version of PhoneGap is also required if you want to deploy your applications directly to the phone instead of running them in the PhoneGap Developer App.

iOS

To deploy PhoneGap apps to iOS devices, you need to install the iOS SDK (*https://developer.apple.com/ios/*) on a Mac. Launch Xcode one time so it can complete the installation after you agree to the license agreement. After Xcode is installed, open a terminal and run xcode-select to install the command-line tools.

```
$ xcode-select --install
```

Use npm to install the deployment tools ios-sim and ios-deploy.

```
$ sudo npm install -g ios-sim
$ sudo npm install -g ios-deploy
```

Refer to the iOS Platform Guide (*http://docs.cordova.io*) for more detailed instructions.

Android

You can develop PhoneGap applications for Android using Windows, Mac, or Linux.

Install the Java SDK (*http://bit.ly/oozie-oracle-java*) on your computer. For Windows, be sure to set the JAVA_HOME environment variable to point to the JDK location. For example, C:\Program Files\Java\jdk1.8.0_45.

Install the Android SDK (*https://developer.android.com/sdk/index.html*). You can choose the standalone version or Android Studio. Follow the installation instructions from Google.

You need to include the tools and platform_tools directories from the Android SDK in your system path. For example, if you installed to /usr/local/android-sdk, add the following to ~/.bash-profile:

```
ANDROID_SDK=/usr/local/android-sdk
export PATH=$PATH:$ANDROID_SDK/tools:$ANDROID_SDK/platform-tools
```

Source bash profile to load the new PATH setting into the current shell.

```
source ~/.bash_profile
```

On Windows, right-click on Computer, select Properties, select Advanced System Settings, and open Environment Variables. Append the SDK location to the PATH variable.

```
;C:\android-sdk\platform-tools;C:\android-sdk\tools
```

Use the Android SDK Manager (shown in Figure 2-6) to install the build tools and the SDK packages. Launch the SDK manager by typing android from a command prompt:

```
$ android
```

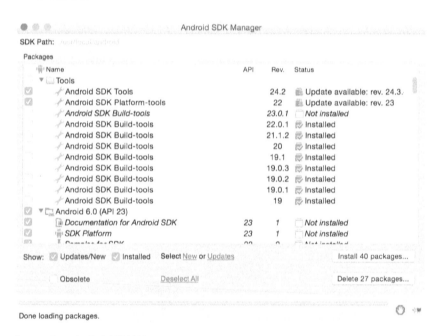

Figure 2-6 *Android SDK Manager*

Refer to the Android Platform Guide (*http://docs.cordova.io*) for more detailed instructions.

Developer Mode

You need to put your Android phone into developer mode before you can deploy applications.

1. Open Settings → System → About Phone.

2. Tap the build number seven times to enable developer mode.

3. Go back one screen, enter developer options, and enable USB debugging.

Smart Light Switch | 3

One of the problems with the current generation of smart lightbulbs is that the smarts are in the bulb.

Though they have been held up as a massive Internet of Things success story, in the long term, this will not turn out to be the case because lightbulbs are something you turn on and off from a switch on the wall. You can make most smart lightbulb systems unresponsive by using the wall switch. Effectively, a smart lightbulb replaces a thing we use every day, the light switch, but it does it poorly. We really need to replace the switch, not the bulb.

What Is a Smart Switch?

A smart light switch not only lets you turn the light on and off using the switch itself but also remotely via Bluetooth LE. The switch should also know its current status—in other words, whether the bulb is on or off—and send a notification over Bluetooth to subscribed applications when the switch is toggled to allow them to update their local status.

Hardware

We need the following hardware to build the light switch:

- Arduino Uno (*http://www.makershed.com/products/arduino-uno-revision-3*)
- Adafruit nRF8001 Bluefruit LE (*https://www.adafruit.com/products/1697*)
- A breadboard
- An LED
- A 220Ω and a 10kΩ resistor
- Jumper wires

- A tactile button switch

Optionally (see "Using Real Lightbulbs") we'll also use a:

- PowerSwitch Tail (*http://www.makershed.com/products/powerswitch-tail-ii*)

The Breadboard

A solderless breadboard (shown in Figure 3-1) is an indispensable tool for rapidly and cheaply prototyping projects.

Figure 3-1 *A typical half-size breadboard*

Breadboards consist of many tiny "holes" arranged on a 0.1-inch grid into which the leads of the component can be connected. Typically, as depicted by red arrows in Figure 3-1, the holes down each side of the breadboard are electrically connected lengthways down the board and are used for the positive and negative (i.e., ground or Earth) power supply. These are usually referred to as *rails*, and will commonly be labeled as such. The other holes in the board are connected horizontally across the board, usually with a gap down the middle. Each hole is connected to the many metal strips that run underneath the board.

Putting the legs of a component in the same row, therefore, forms connections between different components of your circuit. Typically, you'll make use of short lengths of wire, commonly referred to as *jumper wires*, to connect rows of the board.

When using a breadboard, you must use single-core 0.6mm-diameter wire. Stranded wire is not suitable because it will crumple when pushed into a hole, and it may damage the board if individual strands break off.

When using chips with many legs (integrated circuits or, more commonly, ICs), place them in the middle of the board so that half of the legs are on one side of the middle gap that runs down the board and half are on the other side. Since the chip spans the gap in the middle of the board, the legs on one side of the chip will not be electrically connected to the legs on the other side of the chip.

Getting Started

Blinking an LED is the Hello, World of hardware. We are going to blink an LED to ensure your hardware and programming environment are set up correctly.

Resistor Color-Coding

Before you try to wire anything up, you need to get familiar with how resistors are labeled so you can be sure to choose the right ones.

A color code (see Figure 3-2) is normally used to denote the resistance value of a resistor. Normally, there will be four bands: three on one side to tell you what the resistance of the component is, and then a gap, and another band off to the side to tell you the tolerance of the component. In some cases, there may be an additional band as part of the resistor value, but it's fairly rare.

Orient the resistor so that the band that is separated from the others is on the right, and then read the color bands left to right. In the case of our 10kΩ, we see that the bands are brown, black, and orange, which translates to 1, 0 (so 10), and then a multiplier of 1kΩ (hence 10kΩ).

To the right of this is a gold band, which means that our resistor is 10kΩ ± 5%.

Color	Band 1	Band 2	Band 3	Multiplier	Tolerance
Black	0	0	0	1Ω	–
Brown	1	1	1	10Ω	±1%
Red	2	2	2	100Ω	±2%
Orange	3	3	3	1kΩ	
Yellow	4	4	4	10kΩ	
Green	5	5	5	100kΩ	±0.5%
Blue	6	6	6	1MΩ	±0.25%
Violet	7	7	7	10MΩ	±0.1%
Gray	8	8	8		±0.05%
White	9	9	9		
Gold				0.1	±5%
Silver				0.01	±10%

Figure 3-2 *Resistor color-coding chart*

Blinking an LED

Let's go ahead and start building. Grab your Arduino board, a breadboard, an LED, a 220Ω resistor, and some jumper wires.

 It's standard practice to use red wires for power (VCC), which in our case is +5V, and black wires for ground (GND). This helps a lot later on when you're struggling with a breadboard full of components and trying to figure out where all the wires are going.

Connect the +5V pin on your Arduino to the positive rail of the breadboard, and one of the ground (GND) pins to the negative rail. Next, grab the LED and the 220Ω resistor from your pile of parts.

An LED is a light-emitting diode, and diodes only allow current to flow in one direction, so we need to make sure we plug things in correctly. We need the resistor to limit the amount of current that flows through the LED; otherwise, it's liable to burn out.

Most LEDs have a long and a short lead. The longer lead is the positive lead, or *anode*. The negative lead or *cathode* is the shorter lead. Depending on the LED, one side of the LED may be flatter; this is the positive (anode) side.

Take the 220Ω resistor and the LED and wire them up as shown in Figure 3-3, remembering to connect the short leg of the LED toward GND.

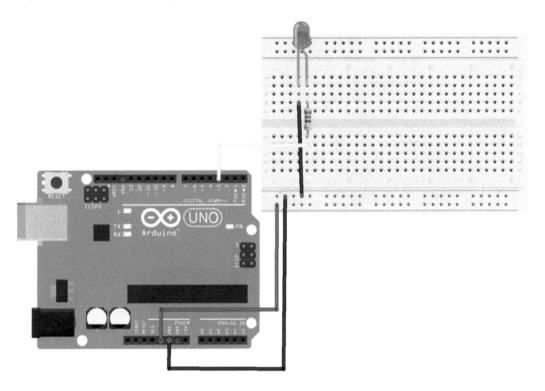

Figure 3-3 *Wiring an LED to our Arduino board*

Now that you've wired your circuit together, go ahead and open the Arduino IDE. There are a number of example sketches included in the development environment, and one of them will do just what we want. The one we're looking for can be found by selecting File → Examples → 1.Basics → Blink, which is shown in Figure 3-4.

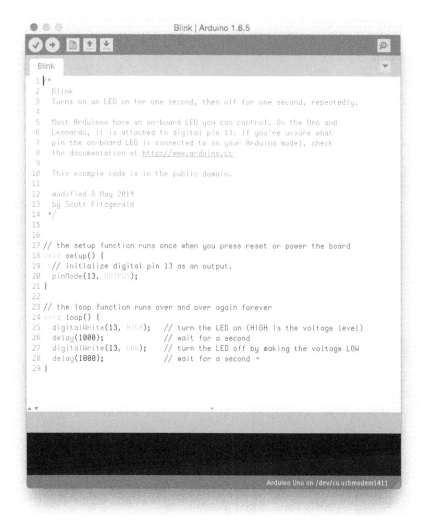

Figure 3-4   *The Blink example sketch*

 An Arduino program is normally referred to as a sketch.

Every Arduino sketch consists of two parts: the setup and the loop. Every time the board is powered up, or the board's reset button is pushed, the setup( ) routine is run. After that finishes, the board runs the loop( ) routine. When that completes, and perhaps somewhat predictably, the loop( ) is run again and again. Effectively, the contents of the loop( ) sit inside an infinite while loop.

Before building and deploying this example to our Arduino, let's take a look at the code:

```
void setup() {
  pinMode(13, OUTPUT);  ❶
}

void loop() {
  digitalWrite(13, HIGH);  ❷
  delay(1000);
  digitalWrite(13, LOW);  ❷
  delay(1000);
}
```

❶ We set pin 13 to behave as an OUTPUT pin. In this state, the pin can provide up to 40 mA of current to other devices. This is enough current to brightly light up an LED, or run many sensors, but not enough current to run most relays, solenoids, or motors.

❷ If the pin has been configured for OUTPUT, its voltage will be set to the corresponding value: 5V for HIGH, 0V (ground) for LOW.

Effectively, then, this piece of code causes the voltage on pin 13 to be brought to HIGH for a second (1,000 ms) and then LOW for a further second before the loop starts again, bringing the voltage to HIGH once more.

As mentioned in Chapter 2, some of the digital pins on the Arduino board have special functions; pin 13 is one of these. On most boards (including the Uno), it has an LED and resistor attached to it that's soldered onto the board itself.

We've attached our LED on the breadboard to pin 3, rather than 13. So go ahead and edit the code, changing all three occurrences of pin number 13 to pin 3.

Make sure you're connected to the Arduino, see "Connecting to the Board", and then go ahead and compile and upload the sketch to your board.

The first step to getting the sketch ready for transfer over to the Arduino is to click on the Verify/Compile button (see Figure 3-5). This will compile your code, checking it for errors, and then translate you program into something that is compatible with the Arduino architecture. After a few seconds, you should see the message Done compiling. in the Status Bar and something along the lines of Binary sketch size: 1,030 bytes of program storage space. Maximum is 32,256 bytes in the Notification area.

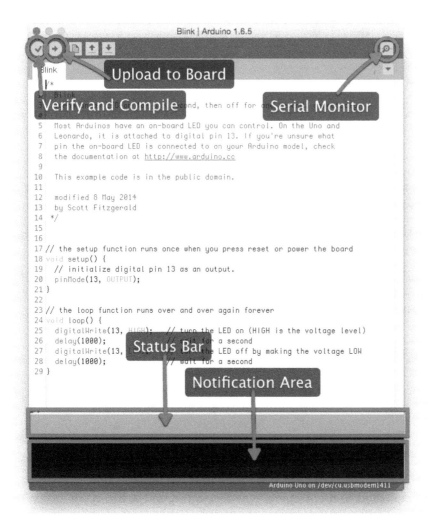

Figure 3-5 *The Arduino development environment*

Once you see that message, go ahead and click on the Upload button (see Figure 3-5 again). This will initiate the transfer of the compiled code to the board via the USB connection.

Wait a few seconds—you should see the RX and TX LEDs on the board flashing as the data is transferred over the serial connection from your computer to the board. If the upload is successful, the message Done uploading. will appear in the status bar.

A few seconds after the upload finishes, you should see the LED start to blink. One second on, one second off. If you see that, congratulations, you've just successfully gotten the hardware equivalent of "Hello, World" to compile and run on the Arduino board.

Adding a Switch

Now that we've seen how a basic Arduino sketch works, let's move on and add a switch. We'll start out with a simple push button, which is called a *momentary contact switch*. These types of switches come in two types—normally open and normally closed, although normally open are by far the most commonly available. Normally open switches complete a circuit (turn things on) when you press them, and then off when you release them.

The most common type of tactile switch intended for breadboarding has four legs, not two as you might expect. The two legs on each side of the switch are connected together. There are really only two wires here, not four. The two switches in Figure 3-6 are therefore the same.

Figure 3-6 *Tactile push-button switches*

Grab the push button and a 10kΩ resistor from your parts pile and wire them up on the breadboard as shown in Figure 3-7.

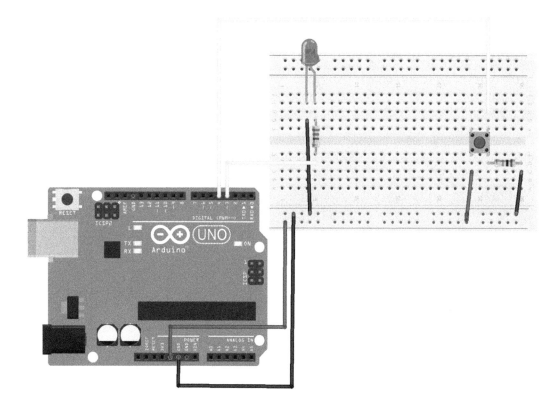

Figure 3-7 *Adding a switch*

What we've done here with the resistor, as you can see in Figure 3-8, is called *debouncing* the button. If we simply connected one side of the button to +5V and the other to pin 4 of the Arduino without a route to GND via the 10kΩ resistor, the pin would indeed be pulled to +5V when we pushed the button. However, when the button was not pushed—and the two sides of the switch were unconnected—then the pin would be *floating* because it's not connected to anything.

Figure 3-8 *How the switch legs are connected*

What that means for us is that our code will work unreliably; sometimes we'll detect button pushes that don't exist. To avoid that, we use the resistor to pull down the pin connection to GND. So when the the switch is not being pushed, the pin is no longer floating; instead it is pulled to a LOW state.

If you've followed the instructions and wired the breadboard as laid out in Figure 3-7, you should have something that looks a lot like Figure 3-9.

Now we're ready to write the software that goes with the hardware. Open a new sketch by clicking on File→New to open a new window.

```
#define LED_PIN      3
#define BUTTON_PIN   4

int val;

void setup() {
  pinMode(LED_PIN, OUTPUT);      ❶
  pinMode(BUTTON_PIN, INPUT);    ❷
}


void loop() {  ❸
  val = digitalRead(BUTTON_PIN);
  if (val == HIGH) {
    digitalWrite(LED_PIN, HIGH);
  }
  if (val == LOW) {
    digitalWrite(LED_PIN, LOW);
  }
}
```

❶ The mode of the pin the LED is connected to is set to OUTPUT, which means we can pull the pin to HIGH or LOW.

❷ The mode of the pin the button is connected to is set to INPUT, which means we can detect whether the pin is being pulled HIGH or LOW.

❸ In the loop we simply check to see if pin 4, the pin the button is connected to, is pulled HIGH or LOW and set the value of the LED accordingly.

Figure 3-9 *The Arduino and breadboarded LED and button*

Save the contents of the sketch to a file using the File→Save menu item, and then click on the Verify button to compile your sketch—and, if all goes well, the Upload button to upload it to the board. You should see the RX and TX LEDs light up as the code is transferred. When you see the `Done uploading.` message, go ahead and push the button. While pushing the button, the LED should go on, and when you release the button, it should go off.

Software Debouncing

In addition to hardware debouncing the button, we can do something called software debouncing to add a little bit more reliability to our code. Open up the Arduino browser and modify your code as shown here:

```
#define LED_PIN       3
#define BUTTON_PIN    4

int currentState;
int debounceState;

void setup() {
  pinMode(LED_PIN, OUTPUT);
  pinMode(BUTTON_PIN, INPUT);
}


void loop() {
  currentState = digitalRead(BUTTON_PIN); ❶
  delay(10);
  debounceState = digitalRead(BUTTON_PIN); ❶

  if ( currentState == debounceState ) { ❷
    if (currentState == HIGH) {
      digitalWrite(LED_PIN, HIGH);
    }
    if (currentState == LOW) {
      digitalWrite(LED_PIN, LOW);
    }
  }
}
```

❶   Read the switch state once and then, 10ms later, we read it again.

❷   If we read the same value twice, only then do we change the LED state.

Here, we read the switch state twice, at a very short interval. If both readings are the same, only then do we go ahead and change the state of the LED. While not strictly necessary, since we've hardware-debounced the button, it's common to do both of these steps for reliability.

If you save your code and hit the Upload button to upload the sketch to the board, and then push and release the button, everything should work as before—but ever so slightly more reliably.

Making a Real Light Switch

Right now our light switch doesn't really behave like a light switch. Our wall switches at home are toggles: flip them one way and the light goes on and stays on; flip them the other way and the lights go off and stay off. They wouldn't be much use if we had to stand and hold the switch down to use them.

We can replicate this by using something called a Single Pole Double Throw (SPDT) switch, as shown in Figure 3-10.

Figure 3-10 *A Single Pole Double Throw (SPDT) switch*

By Changing the Hardware

An SPDT switch has three terminals (A, B, and C). Generally the middle terminal is the "common" terminal and is usually connected to the supply voltage. When the switch is flipped to one side, two of the three terminals come into contact (A and B), and when flipped to the other side, two other terminals come into contact (B and C).

In our case, wiring B up 5V, and A up to the button pin on our Arduino, remembering to pull down terminal A using our 10kΩ resistor and just leaving C floating, would allow us to replicate a normal light switch without making any changes to our code (see Figure 3-11).

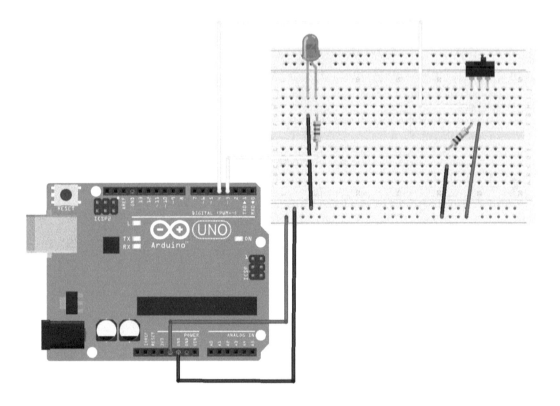

Figure 3-11 *Wiring up the SPDT switch to the Arduino*

If you've followed the instructions and wired the breadboard as laid out in Figure 3-11 the LED should turn on when the SPDT switch is slid to the left, and off when it is slid to the right. No code changes are necessary.

Unfortunately, SPDT switches, especially breadboardable ones, are a lot more expensive than the simple tactile switches we've been working with so far. If you don't have one on hand, or don't want to spend the extra money, then we can also replicate this in code.

Changing the Software

Going back to our push-button switch then, as we had in Figure 3-7, let's open the Arduino development environment and modify our code as follows:

```
#define LED_PIN      3
#define BUTTON_PIN   4

int currentState;
int debounceState;
int switchState = 0;   ❶
int ledState = 0;

void setup() {
```

```
    pinMode(LED_PIN, OUTPUT);
    pinMode(BUTTON_PIN, INPUT);
  }

void loop() {
  currentState = digitalRead(BUTTON_PIN);
  delay(10);
  debounceState = digitalRead(BUTTON_PIN);

  if( currentState == debounceState  ) {
    if ( currentState != switchState ) { ❷

      if ( currentState == LOW ) { ❸
        // Button just released

      } else { ❹
        if ( ledState == 0 ) {
          digitalWrite(LED_PIN, HIGH);
          ledState = 1;
        } else {
          digitalWrite(LED_PIN, LOW);
          ledState = 0;
        }
      }
      switchState = currentState;
    }
  }
}
```

❶ We introduce two state variables: switchState and ledState. When the code is first run, we assume the LED is turned off and the switch is in a "not pressed" state.

❷ We only want to change the LED state when the switch state has changed.

❸ However, if the newly measured state of the LED is LOW, then it must previously have been HIGH. The switch has been pressed, and is now being (or has just been) released. We don't want to do anything at this point.

❹ The only point where we want to toggle the LED state is when the switch state has been changed to HIGH; in other words, the button has just been pressed.

If you save your code and hit the Upload button to upload the sketch to the board, and then push and release the button, the LED should turn on when the button is pushed down and then remain on when you release it. Pushing the button again will turn on the LED, and it should remain off when you release the button.

Adding Bluetooth

Now that we have a working push-button light switch, let's add some Bluetooth LE. We want to be able to turn the light, or LED in this case, on and off using the physical switch but also via Bluetooth LE.

We're going to create a Bluetooth LE peripheral with a single service having two characteristics: a readable/writeable characteristic called "Switch," which is the characteristic we'll use to flip the switch on and off, and a second characteristic called "Status," which we can subscribe to be told about changes in the status of the switch.

We also need descriptors, which are optional, but we can use the Characteristic User Description 0x2901 to provide a text description of the characteristic value. This is a standard UUID, and tools that know about it can use it to give a plain-text description of our service to an end user.

Our service will therefore look something like Table 3-1.

Table 3-1 *Light Switch Service FF10*

Characteristic	UUID	Properties	Comment
Switch	FF11	read, write	1 on, 0 off
State	FF12	notify	1 on, 0 off

But before we go back to our code to implement our service using the BLEPeripheral library we installed in "Installing the BLE Peripheral Library", let's go ahead and wire our Bluetooth LE module to our Arduino.

Wiring Up the Adafruit Bluefruit LE Module

The Bluetooth board we'll be using throughout this book is the Adafruit Bluefruit LE (*http://www.adafruit.com/products/1697*) board based around the Nordic Semiconductor nRF8001 chipset.

You should be careful when purchasing a board, and make sure you're buying the correct one as Adafruit uses the "Bluefruit" name with all of their Bluetooth LE boards as well as some Bluetooth-LE-like products. For instance, their UART Friend (http://www.adafruit.com/products/2479) and SPI Friend (http://www.adafruit.com/products/2633) boards are based around an entirely different chipset but still branded with the "Bluefruit" name.

Wire the board to the Arduino as shown in Figure 3-12, using pin 2 for RDY and pins 9 and 10 for RST and REQ, respectively.

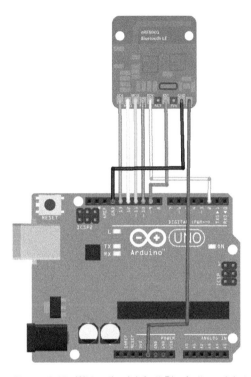

Figure 3-12 *Wiring the Adafruit Bluefruit module to the Arduino*

With the Bluetooth LE module now wired into the Arduino, you should have something that looks like Figure 3-13.

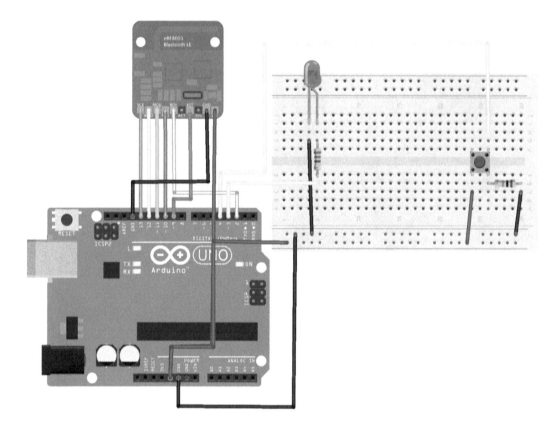

Figure 3-13 *The smart light switch wiring*

Modifying Our Sketch

If you've followed the instructions and wired the breadboard as laid out in Figure 3-13, you should have something that looks a lot like Figure 3-14.

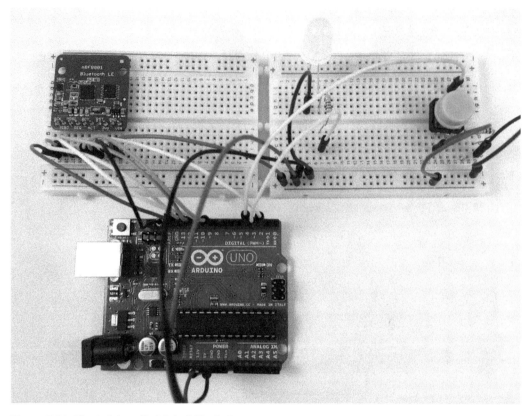

Figure 3-14 *The Arduino with Adafruit Bluefruit*

Now that we have our hardware wired together, let's go back to the development environment and modify the code (shown in Example 3-1) for a push-button light switch.

Example 3-1 *The push-button light switch code*

```
#include <SPI.h>
#include <BLEPeripheral.h>

#define LED_PIN      3
#define BUTTON_PIN   4

#define BLE_REQ     10
#define BLE_RDY      2
#define BLE_RST      9

int currentState;
int debounceState;
int switchState = 0;
int ledState = 0;
```

```
BLEPeripheral blePeripheral = BLEPeripheral(BLE_REQ, BLE_RDY, BLE_RST); ❶
BLEService lightswitch = BLEService("FF10"); ❷

BLECharCharacteristic switchCharacteristic =
BLECharCharacteristic("FF11", BLERead | BLEWrite); ❸
BLEDescriptor switchDescriptor = BLEDescriptor("2901", "Switch");

BLECharCharacteristic stateCharacteristic = BLECharCharacteristic("FF12", BLENotify); ❹
BLEDescriptor stateDescriptor = BLEDescriptor("2901", "State");

void setup() {
  Serial.begin(9600);

  pinMode(LED_PIN, OUTPUT);
  pinMode(BUTTON_PIN, INPUT);

  blePeripheral.setLocalName("Light Switch");  ❺
  blePeripheral.setDeviceName("Smart Light Switch");
  blePeripheral.setAdvertisedServiceUuid(lightswitch.uuid()); ❻

  blePeripheral.addAttribute(lightswitch); ❼
  blePeripheral.addAttribute(switchCharacteristic);
  blePeripheral.addAttribute(switchDescriptor);
  blePeripheral.addAttribute(stateCharacteristic);
  blePeripheral.addAttribute(stateDescriptor);

  blePeripheral.begin(); ❽

  Serial.println(F("Smart Light Switch"));
}

void loop() {
  blePeripheral.poll(); ❾

  currentState = digitalRead(BUTTON_PIN);
  delay(10);
  debounceState = digitalRead(BUTTON_PIN);

  if( currentState == debounceState  ) {
    if ( currentState != switchState ) {

      if ( currentState == LOW ) {
        // Button just released

      } else {
        Serial.print(F("Button event: "));
        if ( ledState == 0 ) {
          stateCharacteristic.setValue(1); ❿
          switchCharacteristic.setValue(1);
          digitalWrite(LED_PIN, HIGH);
          ledState = 1;
          Serial.println(F("light on"));

        } else {
```

```
            stateCharacteristic.setValue(0);  ⓫
            switchCharacteristic.setValue(0);
            digitalWrite(LED_PIN, LOW);
            ledState = 0;
            Serial.println(F("light off"));

        }
      }
      switchState = currentState;
    }
  }
}
```

❶ Here, we create a peripheral instance using the BLEPeripheral library.

❷ Create a service with UUID of 0×FF10.

❸ Create the Switch Read/Write characteristic and descriptor.

❹ Create the State characteristic and descriptor for notifications.

❺ Set the advertised Local Name and Device Name; see "Device Name Versus Local Name" for more details about why we need to set both of these advertised characteristics.

❻ Set the UUID of the advertised service.

❼ Go ahead and add the service and associated characteristics and descriptors to the peripherals instance.

❽ Begin advertising the Bluetooth LE service.

❾ Poll of Bluetooth LE messages.

❿ Set both the switch and state to be *on*.

⓫ Set both the switch and state to be *off*.

If you save your code and hit the Upload button to upload the sketch to the board, you should be able to turn the LED on and off as normal using the button, and if you click on the Serial Console button (see Figure 2-2) in the development environment to open the serial console, you should see the string Smart Light Switch appear, with further messages every time you push the button to turn the LED on or off.

Device Name Versus Local Name

The Device Name is the Device Name 0x2a00 characteristic (*http://bit.ly/gap-device-name*) of the Generic Access 0x1800 service (*http://bit.ly/ generic-access-service*). The Generic Access service is required by the Bluetooth GATT Spec. If you don't set a Device Name, the BLEPeripheral will use a default value of "Arduino".

The Local Name is the name that is advertised by BLEPeripheral in the scan data. This is part of the Bluetooth Generic Access Profile, or GAP.

We need to set both the Device Name and the Local Name. Tools like LightBlue on iOS get the Device Name from the Generic Access service. Tools like nRF Master Control Panel on Android rely on the name in the advertising packet.

However, right now we can read, but we can't write (change) the state of the LED. In other words, right now we can't "throw" the switch using Bluetooth LE. We need a handler foundation. Just above the `blePeripheral.begin( )`, add the following:

```
switchCharacteristic.setEventHandler(BLEWritten, switchCharacteristicWritten); ❶
```

❶   Assign an event handler method to be called when a write command is made on the peripheral.

Then, at the bottom of your sketch, after the `loop( )` function, add the handler function itself.

```
void switchCharacteristicWritten(BLECentral& central, BLECharacteristic& characteristic)
{
  Serial.print(F("Characteristic event: "));
  if (switchCharacteristic.value()) {
    Serial.println(F("light on"));
    digitalWrite(LED_PIN, HIGH);
    ledState = 1;
    stateCharacteristic.setValue(1);

  } else {
    Serial.println(F("light off"));
    digitalWrite(LED_PIN, LOW);
    ledState = 0;
    stateCharacteristic.setValue(0);

  }
}
```

If you save your code and hit the Upload button to upload the sketch to the board, everything should continue to work as before. However, now you have everything in place to control the LED, not just with the button, but via Bluetooth LE. What we need is a generic Bluetooth LE explorer application so we can easily examine and trigger our service.

Testing the Service

If you're an iOS user, go ahead and install LightBlue (*http://bit.ly/lightblue-ble*) on your iPhone or iPad. Alternatively if you're an Android user, install nRF Master Control Panel (*http://bit.ly/nRF-ble*) on your phone or tablet. The two apps present the same information but do so somewhat differently.

Opening either app will start it scanning for Bluetooth LE devices. You'll be able to choose a peripheral from a list of nearby devices and explore information about that connected peripheral, its services, and characteristics.

Taking a look at our Smart Light Switch in LightBlue you should see something along the lines of Figure 3-15.

Figure 3-15 *Exploring the Smart Light Switch in the LightBlue app*

Tapping through from the Smart Light Switch in the peripherals list, you can see the advertisement data for the service showing our two chracteristics: Switch, which is Read/Write, and State, which is Notify. We can register the LightBlue app for notifications when the LED state changes by tapping on the Listen for Notifications button in the State characteristic screen.

Going back, we can then go to the Switch characteristic screen, which shows the current value of the chractersitic, which should be 0x00, meaning the LED is off. Tap on "Write new value" to open the editor. Enter 01 and hit Done; the LED should turn on and the characteristic screen should show the new value as 0x01. If you registered for notifications, you should see a canvas drop-down (see Figure 3-16) to tell you the value has changed.

Figure 3-16 *We are notified that the state of the LED has changed*

If you have the Serial Console open you should also see a `Characteristic event: light on` message printed in the console. Finally, if you push the button, you should see a further notification in LightBlue that the LED state has changed back to 0x00.

If everything works, that's it, we have a working smart light switch.

Using Real Lightbulbs

Right now, our light switch is just a proof of concept; turning an LED on and off isn't actually as interesting as it looks at first. But it's actually pretty easy to turn our simple breadboarded project into one that can turn a real light on and off.

Here's what you'll need:

- A PowerSwitch Tail (*http://www.makershed.com/products/powerswitch-tail-ii*)

The PowerSwitch Tail (see Figure 3-17) simplifies our lives by hiding all that nasty AC electricity and letting us use a relay and our Arduino board to turn real mains-powered devices on and off.

Figure 3-17 *The PowerSwitch Tail*

On the side of the PowerSwitch Tail, you will need to connect three wires to the terminal block. Use a small watchmaker's screwdriver to access the screws from the top of the Tail. Turn the screws counterclockwise to open the terminal contacts, and insert the wires into the terminal block contacts through the holes on the side of the Tail.

The left-most contact (labeled +in) is for +5V; the middle wire (labeled -in) is the signal wire; and the right-most wire (labeled Ground) is, as you'd expect, GND.

We then need to make some changes to our hardware. Go ahead and wire up the Arduino, switch, and Powerswitch Tail as shown in Figure 3-18.

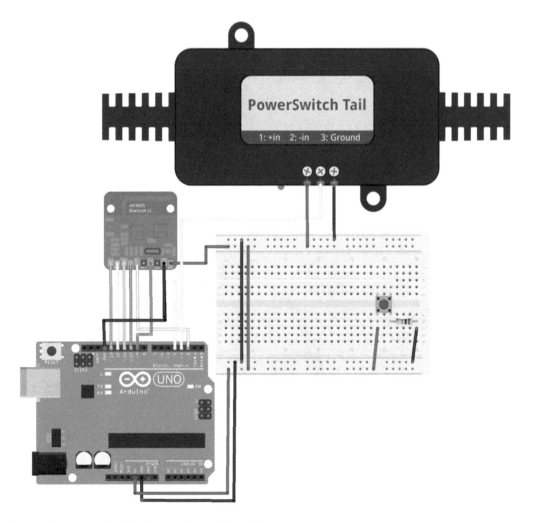

Figure 3-18 *Swapping the LED for a PowerSwitch Tail*

Then plug the PowerSwitch Tail into the wall, and then plug in a mains-powered lamp or any electrical device (maximum draw is 15amps at 120V) that you would like to control to the PowerSwitch Tail socket.

If you've followed the instructions and wired the breadboard as laid out in Figure 3-18, you should have something that looks a lot like Figure 3-19 at this point.

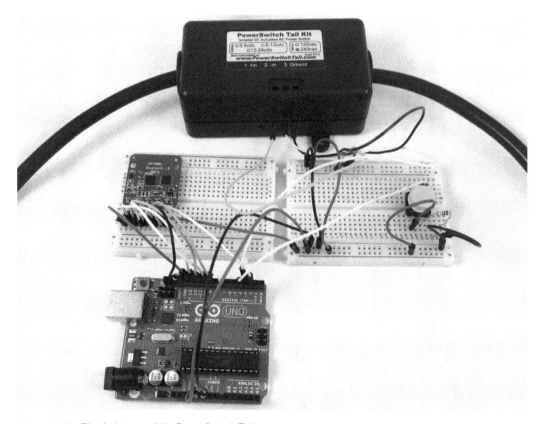

Figure 3-19 *The Arduino and the PowerSwitch Tail*

The Powerswitch Tail can be wired either as "normally open" or "normally closed." In the normally open configuration, when the signal (middle) terminal is pulled and held LOW, then the relay is tripped and power will flow from the mains to the attached lamp. Power will only flow while the signal wire is pulled LOW. Conversely, in the normally closed configuration when the signal terminal is pulled and held LOW, then the relay is tripped and power will cease to flow from the mains to the attached lamp.

Effectively, the default state of the attached mains-powered lamp is off if the Powerswitch Tail is normally open, and on for if it is normally closed. The most common way to wire the Powerswitch Tail is "normally open," which makes sense from a safety perspective. If there isn't a signal from the Arduino, then the attached mains device is "off."

Since pulling the signal wire LOW rather than HIGH is what triggers the relay, in the "normally open" case we have to flip the logic in our code. Go back into the code and everywhere there is a

```
digitalWrite(LED_PIN, HIGH);
```

change it to

```
digitalWrite(LED_PIN, LOW);
```

and vice versa. In other words, we need to swap the logic around for the LED_PIN, changing HIGH to LOW and LOW to HIGH for the "normally open" case.

Conversely, in the "normally closed" case, the attached lamp will be on when the LED_PIN is pulled HIGH, and then will turn off when we pull the pin LOW, so no change to our code is needed from when we were working with the LED.

 If you swap the LED out for the Powerswitch Tail and the lamp behaves in the opposite manner of what you're expecting—it turns on when you think it should turn off, and off when you think it should turn on—then just flip the values for the LED_PIN in the code.

If you save your code and hit the Upload button to upload the sketch to the board, everything should continue to work as before. Except now, instead of controlling an LED, you're controlling a real lamp.

Conclusion

In this chapter, we used a Arduino to build a smart light switch that can control a real-world device using the PowerSwitch Tail.

BLE Lock

4

In this chapter you'll build a lock that can be opened using your phone. The lock mechanism will be built with an Arduino Uno, Bluetooth LE radio, and a solenoid lock. You'll also write an iOS and Android application using PhoneGap that opens the lock using Bluetooth LE.

Lock Service

Most Bluetooth LE hardware for Arduino comes with a service that emulates the Bluetooth classic Serial Port Profile (SPP) (*http://bit.ly/1SaYSU2*). While an SPP-like service could work for this project, it doesn't take full advantage of Bluetooth LE. For this project, use the custom lock service defined in Table 4-1.

Table 4-1 *Lock Service D270*

Characteristic	UUID	Properties
unlock	D271	write
message	D272	notify

The lock service does two things. The unlock characteristic allows the lock to be opened by writing a secret code to the characteristic. The status message characteristic allows the lock to send information, such as `invalid code` or `unlocked`, back to the client. If a Bluetooth client subscribes to notifications on the message characteristic, it will be automatically notified whenever a new message is available.

Hardware

You will need the following hardware to build the lock:

- Arduino Uno
- Adafruit Bluefruit LE nRF8001 (*http://adafru.it/1697*) or RedBearLab BLE Shield (*http://bit.ly/1NRAYLJ*)
- Solenoid lock (*http://adafru.it/1512*)
- 12 V, 1A power supply
- Darlington NPN Transistor (TIP120)
- Red LED
- Green LED
- Two 220 Ohm resistors

The lock for this project has a solenoid actuated slug. The lock is normally in the closed or locked position. When power is applied to the solenoid, the bolt is drawn back with an electromagnet opening the lock. The lock draws 650mA at 12V, which is more current than you can safely pull through most Arduino pins. Arduino pins have a recommended maximum of 20 mA. The hardware uses pin 6 to switch a Darlington transistor on, sending the higher voltage and current from the VIN pin to the solenoid.

 You might be able to run this project with a 9V power adapter. Some solenoid locks work with 9V, but others don't. I've had better luck with 12V adapters.

The Arduino can handle 12V input because it has internal voltage regulators that power the board and Atmel chip at 5V. The VIN pin allow you to draw 12V current directly from the barrel jack. The 650mA the solenoid lock draws is less than the max of 1A for the VIN pin. Note that the VIN pin can handle much more current than a pin that is connected to the microcontroller.

This project can use either the Adafruit Bluefruit LE nRF8001 breakout board (Figure 4-1) or the RedBearLab BLE Shield (Figure 4-2). Some of the wiring will be different depending on which hardware you choose (see Figure 4-3). The software will be mostly the same since we're using the Arduino BLE Peripheral library to create custom services.

Red and green LEDs are used to show the lock status. The green LED will light when the unlock code is correct. The red LED will light when the code is incorrect. Both the LEDs and the lock reset after four seconds.

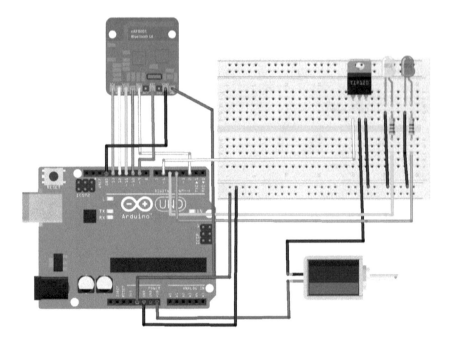

Figure 4-1 *Lock wiring for Adafruit Bluefruit LE nRF8001*

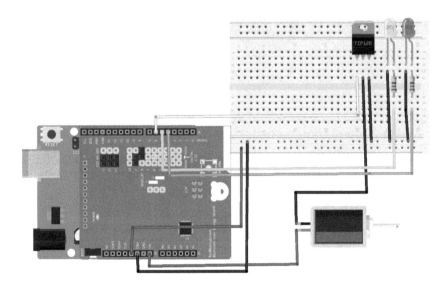

Figure 4-2 *Lock wiring for RedBearLab BLE shield*

Figure 4-3 *Lock wiring for Adafruit nRF8001 on left, RedBearLab BLE shield on right*

Lock Software

The Arduino software hardcodes the passcode used to open the lock. For now, you'll use the password 12345. I expect you to choose a better passcode for your top-secret project.

Use the Arduino IDE to program the hardware. Refer back to "Installing the Arduino IDE" if you need help configuring your development environment. "Installing the BLE Peripheral Library" describes how to install the Arduino BLE Peripheral library.

Programming

Open the Arduino IDE and create a new sketch using File → New. Save the file as *BLE_Lock*.

At the beginning of the sketch, include the SPI and Bluetooth libraries.

```
#include <SPI.h>
#include <BLEPeripheral.h>
```

Define the pins for the lock and LEDs. The pin numbers defined in the code should match where the wires are plugged into the Arduino.

```
#define LOCK_PIN 6
#define RED_LED_PIN 4
#define GREEN_LED_PIN 5
```

Define the pins for the Bluetooth hardware you're using.

For the Adafruit nRF8001, use the definitions in Example 4-1.

Example 4-1 Defines for Adafruit Bluefruit LE

```
// Adafruit Bluefruit LE
#define BLE_REQ 10
#define BLE_RDY 2
#define BLE_RST 9
```

For the RedBearLab BLE Shield, use the settings in Example 4-2.

Example 4-2 Defines for RedBearLab BLE Shield

```
// RedBear BLE Shield 2.x
#define BLE_REQ 9
#define BLE_RDY 8
#define BLE_RST UNUSED
```

You need to translate Table 4-1 into code. Create the BLE peripheral, service, characteristics, and descriptors. The descriptors aren't required but help make the service self-describing.

```
BLEPeripheral blePeripheral = BLEPeripheral(BLE_REQ, BLE_RDY, BLE_RST);
BLEService lockService = BLEService("D270");
BLECharacteristic unlockCharacteristic = BLECharacteristic("D271", BLEWrite, 20);
BLEDescriptor unlockDescriptor = BLEDescriptor("2901", "Unlock");
BLECharacteristic statusCharacteristic = BLECharacteristic("D272", BLENotify, 20);
BLEDescriptor statusDescriptor = BLEDescriptor("2901", "Status Message");
```

The code that opens the lock is stored in a character array.

```
char secret[] = "12345";
```

Add an additional variable to track the last time the lock was opened.

```
long openTime = 0;
```

Setup

Create the setup function. Initialize the serial output so debug information can be sent to the serial monitor.

```
void setup() {
  Serial.begin(9600);
  Serial.println(F("BLE Lock"));
```

You created the blePeripheral instance in the preamble of the sketch. The peripheral needs to be configured in the setup function. Set the device and local name to BLE Lock. Set the lock service UUID as the advertised service UUID.

```
blePeripheral.setDeviceName("BLE Lock");
blePeripheral.setLocalName("BLE Lock");
blePeripheral.setAdvertisedServiceUuid(lockService.uuid());
```

See "Device Name Versus Local Name" for more information about device name and local name.

Add the lock service, characteristics, and descriptors to the peripheral. The order in which the items are added is important. Add the service first, followed by the unlock characteristic and the optional descriptor. Repeat the process for the status message characteristic and descriptor.

```
blePeripheral.addAttribute(lockService);

blePeripheral.addAttribute(unlockCharacteristic);
blePeripheral.addAttribute(unlockDescriptor);

blePeripheral.addAttribute(statusCharacteristic);
blePeripheral.addAttribute(statusDescriptor);
```

When a Bluetooth client writes a new value to the unlock characteristic, the sketch needs to process the data. Add an event handler for the BLEWritten event to the unlockCharac teristic. The unlockCharacteristicWritten function will be defined later.

```
unlockCharacteristic.setEventHandler(BLEWritten, unlockCharacteristicWritten);
```

Now that the blePeripheral is configured, it can be started.

```
blePeripheral.begin();
```

Finally, set the lock and LED pins to OUTPUT and the values to LOW, meaning off.

```
pinMode(LOCK_PIN, OUTPUT);
pinMode(RED_LED_PIN, OUTPUT);
pinMode(GREEN_LED_PIN, OUTPUT);
digitalWrite(LOCK_PIN, LOW);
digitalWrite(RED_LED_PIN, LOW);
digitalWrite(GREEN_LED_PIN, LOW);
} // end of setup function
```

Loop

The loop function has two responsibilities. First, it tells the Bluetooth radio to do whatever it should be working on. Next, it checks the last time someone attempted to open the lock. If the open time is not zero, and was at least four seconds, the lock is closed and the lights are reset.

```
void loop() {

    // Tell the bluetooth radio to do whatever it should be working on
```

```
    blePeripheral.poll();

    // close lock and reset lights after 4 seconds
    if (openTime && millis() - openTime > 4000) {
      resetLock();
    }
```

Unlock Characteristic Written

The implementation of unlockCharacteristicWritten must match the callback signature void bleCharacteristicEventHandler(BLECentral& central, BLECharacteristic& characteristic) (*http://bit.ly/1MXzxYS*). Log the fact that the characteristic changed. Pass the characteristic data to the openLock function.

```
    void unlockCharacteristicWritten(BLECentral& central, BLECharacteristic& characteristic)
    {
      // central wrote new value to the unlock characteristic
      Serial.println(F("Unlock characteristic written"));

      openLock(characteristic.value(), characteristic.valueLength());
    }
```

Open Lock

The open lock function checks the passcode written to the unlock characteristic against the secret code. If the code matches, the green LED is turned on, the lock is opened, and the status message is set to unlocked. If the code does not match, the red LED is turned on and the status message is set to invalid code.

```
    void openLock(const unsigned char* code, int codeLength) {
      openTime = millis();  // set even if bad code so we can reset the lights

      // does the code match the secret
      boolean match = false;

      if (strlen(secret) == codeLength) {
        for (int i = 0; i < codeLength; i++) {
          if (secret[i] != code[i]) {
            match = false;
            break;
          } else {
            match = true;
          }
        }
      }

      if (match) {
        // open the lock
        Serial.println("Code matches, opening lock");
        digitalWrite(GREEN_LED_PIN, HIGH);
        digitalWrite(RED_LED_PIN, LOW);
        digitalWrite(LOCK_PIN, HIGH); // open the lock
        statusCharacteristic.setValue("unlocked");
      } else {
```

```
      // bad code, don't open
      Serial.println("Invalid code");
      digitalWrite(RED_LED_PIN, HIGH);
      statusCharacteristic.setValue("invalid code");
   }
 }
```

Reset Lock

The resetLock function is called from the loop function. It turns off the red and green LEDs, closes the lock, sets the status message characteristic to locked, and resets the open time:

```
void resetLock() {
  // reset the lights
  digitalWrite(RED_LED_PIN, LOW);
  digitalWrite(GREEN_LED_PIN, LOW);
  digitalWrite(LOCK_PIN, LOW); // close the lock
  statusCharacteristic.setValue("locked");
  openTime = 0;
}
```

See Example 4-3 for the complete listing of *BLE_Lock*.

Example 4-3 *BLE_Lock.ino*

```
#include <SPI.h>
#include <BLEPeripheral.h>

#define LOCK_PIN 6
#define RED_LED_PIN 4
#define GREEN_LED_PIN 5

// See BLE Peripheral documentation for setting up your hardware
// https://github.com/sandeepmistry/arduino-BLEPeripheral#pinouts

// Adafruit Bluefruit LE
#define BLE_REQ 10
#define BLE_RDY 2
#define BLE_RST 9

// RedBear BLE Shield 2.x
//#define BLE_REQ 9
//#define BLE_RDY 8
//#define BLE_RST UNUSED

BLEPeripheral blePeripheral = BLEPeripheral(BLE_REQ, BLE_RDY, BLE_RST);
BLEService lockService = BLEService("D270");
BLECharacteristic unlockCharacteristic = BLECharacteristic("D271", BLEWrite, 20);
BLEDescriptor unlockDescriptor = BLEDescriptor("2901", "Unlock");
BLECharacteristic statusCharacteristic = BLECharacteristic("D272", BLENotify, 20);
BLEDescriptor statusDescriptor = BLEDescriptor("2901", "Status Message");
```

```
// code that opens the lock
char secret[] = "12345";
long openTime = 0;

void setup() {
  Serial.begin(9600);
  Serial.println(F("BLE Lock"));

  // set advertised name and service
  blePeripheral.setDeviceName("BLE Lock");
  blePeripheral.setLocalName("BLE Lock");
  blePeripheral.setAdvertisedServiceUuid(lockService.uuid());

  // add service and characteristic
  blePeripheral.addAttribute(lockService);

  blePeripheral.addAttribute(unlockCharacteristic);
  blePeripheral.addAttribute(unlockDescriptor);

  blePeripheral.addAttribute(statusCharacteristic);
  blePeripheral.addAttribute(statusDescriptor);

  // assign event handlers for characteristic
  unlockCharacteristic.setEventHandler(BLEWritten, unlockCharacteristicWritten);

  // begin initialization
  blePeripheral.begin();

  pinMode(LOCK_PIN, OUTPUT);
  pinMode(RED_LED_PIN, OUTPUT);
  pinMode(GREEN_LED_PIN, OUTPUT);
  digitalWrite(LOCK_PIN, LOW);
  digitalWrite(RED_LED_PIN, LOW);
  digitalWrite(GREEN_LED_PIN, LOW);
}

void loop() {
  // Tell the bluetooth radio to do whatever it should be working on
  blePeripheral.poll();

  // close lock and reset lights after 4 seconds
  if (openTime && millis() - openTime > 4000) {
    resetLock();
  }
}

void unlockCharacteristicWritten(BLECentral& central,
                                 BLECharacteristic& characteristic) {
    // central wrote new value to the unlock characteristic
    Serial.println(F("Unlock characteristic written"));
    openLock(characteristic.value(), characteristic.valueLength());
}

void openLock(const unsigned char* code, int codeLength) {
```

```
    openTime = millis();  // set even if bad code so we can reset the lights

    // does the code match the secret?
    boolean match = false;

    if (strlen(secret) == codeLength) {
      for (int i = 0; i < codeLength; i++) {
        if (secret[i] != code[i]) {
          match = false;
          break;
        } else {
          match = true;
        }
      }
    }

    if (match) {
      // open the lock
      Serial.println("Code matches, opening lock");
      digitalWrite(GREEN_LED_PIN, HIGH);
      digitalWrite(RED_LED_PIN, LOW);
      digitalWrite(LOCK_PIN, HIGH); // open the lock
      statusCharacteristic.setValue("unlocked");
    } else {
      // bad code, don't open
      Serial.println("Invalid code");
      digitalWrite(RED_LED_PIN, HIGH);
      statusCharacteristic.setValue("invalid code");
    }
  }

// closes the lock and resets the lights
void resetLock() {
  // reset the lights
  digitalWrite(RED_LED_PIN, LOW);
  digitalWrite(GREEN_LED_PIN, LOW);
  digitalWrite(LOCK_PIN, LOW); // close the lock
  statusCharacteristic.setValue("locked");
  openTime = 0;
}
```

Testing the Lock

Now that the hardware has been built and programmed, you can use a generic Bluetooth application to test the service. Use LightBlue (*http://bit.ly/1hq3m9j*) if you have an iPhone, iPad or iPod. Use nRF Master Control Panel (*http://bit.ly/1Sb9ySu*) if you have an Android device.

iOS

On iOS, use the LightBlue application to connect to the lock (Figure 4-4).

Figure 4-4   *Left: LightBlue connected to the Lock Service; right: LightBlue opening the lock*

1. Select the Status Message characteristic. LightBlue defaults to Hex for displaying characteristic data.

2. Switch the view from Hex to String by selecting Hex from the top-right corner of the screen.

3. Choose UTF-8 String from the list.

4. After the application navigates back to the characteristic view, select the "listen for notifications" link.

5. Use the link on the top-left to navigate back to the peripheral view.

6. Choose the Unlock characteristic. Follow the same process to switch from Hex to String.

7. Touch Hex link on the top-right.

8. Choose UTF-8 String. Now you are ready to open the lock.

9. Touch "Write new value", enter **12345** into the form, and press Done.

10. If you entered the correct code, the lock will open and LightBlue will receive the status notification (Figure 4-4).

Android

Android users should use the nRF Master Control Panel to connect to the lock. The lock service has the 16-bit UUID D270, but the application will display this as the expanded 128-bit version.

1. Choose 0000-**d270**-0000-1000-8000-00805f9b34fb.

2. Subscribe to the Message characteristic by pressing the button with the down arrows next to UUID 0000-**d272**-0000-1000-8000-00805f9b34fb. Now you are ready to send the unlock code to the lock.

3. Click the up arrow next to the Unlock characteristic 0000-**d271**-0000-1000-8000-00805f9b34fb. A new screen will pop up allowing you to write a value.

4. Enter **12345** as the value.

5. Change the BYTE ARRAY drop-down to TEXT.

6. Press the Send button. If you entered the correct code, the lock will open and the nRF Master Control Panel will receive the status notification. Note that the bytes (0x)75-6E-6C-6F-63-6B-65-64 are also displayed as the string "unlocked" (Figure 4-5).

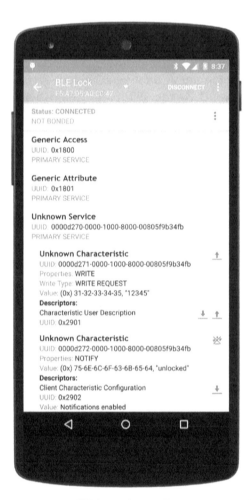

Figure 4-5 *nRF Master Control Panel opening the lock*

Mobile Application

Now that the hardware is set up and you know that BLE works, it is time to write a mobile application that opens the lock. The application will be written using PhoneGap so that one application can run on both iOS and Android.

PhoneGap applications are written using HTML, CSS, and JavaScript, which is packaged into a native iOS or Android application. PhoneGap provides Javascript APIs to access native phone functions like Camera, Contacts, and File System. Support for Bluetooth Low Energy is added with a third-party plugin (*http://github.com/don/cordova-plugin-ble-central*).

Although PhoneGap applications are cross-platform, it requires that the native SDKs for each platform are installed on your system. We are going to circumvent that restriction by using the PhoneGap Developer App (*http://app.phonegap.com*). The PhoneGap Developer

App is a downloaded onto your iOS or Android device from the appstore. A minimal PhoneGap installation on your computer can deploy PhoneGap apps to the developer app running on your phone. This setup is great for experimentation and development, but can not be used to distribute applications. For more information on PhoneGap and getting your system setup refer to "Installing PhoneGap". If you'd like to publish applications, you'll need to install the iOS and Android SDKs in addition to the PhoneGap.

Callbacks

When we're writing JavaScript apps, it's important to understand that many API calls are asynchronous and there are lots of callbacks. This is particularly true with the PhoneGap APIs.

In many programming environments, you call a method and get a result back right away.

```
try {
    BigDecimal temperature = api.getTemperature();
    System.out.println(temperature);
} catch (IOException e) {
    System.out.println("Error getting temperature: " + e);
}
```

PhoneGap is a bit different. You call a method and pass in two functions. The first function is called with a result if the function is successful, and the second function is called if there is an error. The callback functions are invoked asynchronously. Sometimes they appear to be invoked immediately. Other times, callbacks are executed at some point in the future.

```
var success = function(temperature) {
    console.log(temperature);
}

var failure = function(reason) {
    alert("Failed to get temperature: " + reason);
}

api.getTemperature(success, failure);
```

BLE Lock App

Here's the basic process for the mobile application. The application scans to find any hardware advertising the Lock Service. The user selects a device from the list of discovered devices. (In our case, there should only be one lock device listed.) The application connects to the selected peripheral and registers to receive notifications for the message characteristic. Registering for notifications allows the phone to be notified anytime the content of the message characteristic changes. After a successful connection, the user interface is changed to the lock details page so the unlock code can be entered using the number pad.

Open a terminal or cmd prompt on your computer. Change to the directory where you'd like to create the project.

```
cd ~/bluetooth
```

Use the `phonegap` command-line tool to create a new project.

```
phonegap create ble-lock com.makebluetooth.blelock "BLE Lock" --template blank
```

Open the project in your text editor.

CSS

A CSS file is used to style the application and make it look a little better. Create a new directory named `css` inside the `www` folder. Create a new file named `index.css` in the new `css` folder. Copy the premade CSS from GitHub (*http://bit.ly/1SaYzlU*) into *index.css*.

HTML

Open `index.html`. In the `head` section of the file, change title from `Blank App` to `BLE Lock`. Just above the title, add a new line to link to the stylesheet you copied into the project.

```
<link rel="stylesheet" type="text/css" href="css/index.css" />
<title>BLE Lock</title>
```

The user interface of the application is built with HTML. There are two screens in the application. Each screen is an HTML div. Divs are shown and hidden to switch screens. The status div is always visible to display messages to the user. The scrim div is shown while the application is processing data.

The main screen has an unordered list that we populate with the discovered BLE devices. When the user taps a device in this list, the application connects to the lock and switches to the second screen. The second screen has a text field where the user can enter the passcode.

Create the HTML for the device list screen inside the <body> tag of `index.html`.

```
<div id="deviceListScreen">
    <h1>Devices</h1>
    <ul id="deviceList">
        <li>Searching...</li>
    </ul>
    <div>
        <button id="refreshButton">Refresh</button>
    </div>
</div>
```

The unlock screen is shown when the application is connected to a device. The form allows the user to enter the secret code that will be sent to the lock via Bluetooth.

```
<div id="unlockScreen">
    <form>
        <input type="tel" name="code">
        <input type="submit" value="Unlock">
    </form>
    <button id="disconnectButton">Disconnect</button>
</div>
```

The `statusDiv` sits at the bottom of the screen and can display messages to the user.

```
<div id="statusDiv"></div>
```

The scrim div is shown over the user interface while the application is processing data.

```
<div id="scrim">
    <p>Processing</p>
</div>
```

Lastly, the JavaScript files need to be included. PhoneGap automatically adds cordova.js when deploying the application. You will create js/index.js later.

```
<script type="text/javascript" src="cordova.js"></script>
<script type="text/javascript" src="js/index.js"></script>
```

See Example 4-4 for the complete listing of *index.html*.

Example 4-4 *Contents of the index.html file*

```
<!DOCTYPE html>
<html>
    <head>
        <meta charset="utf-8">
        <meta name="viewport" content="initial-scale=1, maximum-scale=1,
         user-scalable=no, width=device-width">
        <link rel="stylesheet" type="text/css" href="css/index.css" />
        <title>BLE Lock</title>
    </head>
    <body>
        <div id="scrim">
            <p>Processing</p>
        </div>
        <div id="deviceListScreen">
            <h1>Devices</h1>
            <ul id="deviceList">
                <li>Searching...</li>
            </ul>
            <div>
                <button id="refreshButton">Refresh</button>
            </div>
        </div>
        <div id="unlockScreen">
            <form>
                <input type="tel" name="code">
                <input type="submit" value="Unlock">
            </form>
            <button id="disconnectButton">Disconnect</button>
        </div>
        <div id="statusDiv"></div>
        <script type="text/javascript" src="cordova.js"></script>
        <script type="text/javascript" src="js/index.js"></script>
    </body>
</html>
```

JavaScript

JavaScript is used to control the application and manipulate the user interface. Create a new directory named js inside the www folder. Create a new file named index.js in the js folder.

At the beginning of index.js file, define some variables with the UUIDs for the Bluetooth Lock Service. These UUIDs are the same ones we used in the Arduino code and are defined in Table 4-1.

```
var SERVICE_UUID = 'D270';
var UNLOCK_UUID = 'D271';
var MESSAGE_UUID = 'D272';
```

The unlock code is entered into the application as a String, but the Bluetooth APIs expect to receive ArrayBuffers of bytes. Create a helper function to convert Strings into ArrayBuffers.

```
function stringToArrayBuffer(str) {
    // assuming 8 bit bytes
    var ret = new Uint8Array(str.length);
    for (var i = 0; i < str.length; i++) {
        ret[i] = str.charCodeAt(i);
        console.log(ret[i]);
    }
    return ret.buffer;
}
```

The status messages from the lock are received as ArrayBuffers. Create a helper method to convert an ArrayBuffer to a String.

```
function bytesToString(buffer) {
    return String.fromCharCode.apply(null, new Uint8Array(buffer));
}
```

Create a variable named app. The application logic is created by adding functions inside the app object.

```
var app = {
};
```

The PhoneGap application is started by calling app.initialize(). Typically, PhoneGap applications call initialize, bindEvents, and onDeviceReady as part of the bootstrapping process. The `initialize` function calls `bindEvents` and hides the divs containing the screens.

```
initialize: function() {
    this.bindEvents();
    deviceListScreen.hidden = true;
    unlockScreen.hidden = true;
},
```

The `bindEvents` function is where HTML DOM events are connected to the application's JavaScript. PhoneGap fires the `deviceready` event once the framework is initialized. It is important for PhoneGap apps to wait for `deviceready` before calling any PhoneGap APIs.

The touch events from buttons and list elements are attached to functions in this object. For example, `scan` is called when the refresh button is tapped.

```
bindEvents: function() {
    document.addEventListener('deviceready', this.onDeviceReady, false);
    refreshButton.ontouchstart = this.scan;
    deviceList.ontouchstart = this.connect;
    document.forms[0].addEventListener('submit', this.unlock, false);
    disconnectButton.onclick = this.disconnect;
},
```

When `deviceready` is fired, the app can begin scanning for Bluetooth peripherals.

```
onDeviceReady: function() {
    app.scan();
},
```

Scan clears any existing items out of the `deviceList` and shows the scrim div with a message before scanning for devices.

The UUID of the Lock Service is passed to the `ble.startScan` function. This limits discovery to only those BLE peripherals that are advertising the Lock Service. The success callback `app.onDeviceDiscovered` is called *every time* a Bluetooth peripheral is discovered. The failure callback is an inline function that shows an alert if something goes wrong.

```
scan: function(e) {
    deviceList.innerHTML = "";  // clear the list
    app.showProgressIndicator("Scanning for Bluetooth Devices...");

    ble.startScan([SERVICE_UUID],
        app.onDeviceDiscovered,
        function() { alert("Listing Bluetooth Devices Failed"); }
    );

    // stop scan after 5 seconds
    setTimeout(ble.stopScan, 5000, app.onScanComplete);

},
```

The function `onDeviceDiscovered` is called each time a peripheral is found. It receives a JSON description of the peripheral object. The `onDeviceDiscovered` function ensures the `deviceListPage` is visible, builds a new list item element with the peripheral information, and adds the new element to the device list.

The device's id is stored in the list item's dataset. This allows the device id to be easily retrieved when a list item is selected.

```
onDeviceDiscovered: function(device) {
    var listItem, rssi;

    app.showDeviceListScreen();

    console.log(JSON.stringify(device));
    listItem = document.createElement('li');
    listItem.dataset.deviceId = device.id;
    if (device.rssi) {
        rssi = "RSSI: " + device.rssi + "<br/>";
    } else {
        rssi = "";
    }
    listItem.innerHTML = device.name + "<br/>" + rssi + device.id;
    deviceList.appendChild(listItem);

    var deviceListLength = deviceList.getElementsByTagName('li').length;
    app.setStatus("Found " + deviceListLength +
                " device" + (deviceListLength === 1 ? "." : "s."));
},
```

A timer in the `scan` method calls `ble.stopScan` after five seconds. The `onScanComplete` function is the success callback for `ble.stopScan`. The `onScanComplete` function checks if the list of discovered devices is empty and sets a status message to notify the user.

```
onScanComplete: function() {
    var deviceListLength = deviceList.getElementsByTagName('li').length;
    if (deviceListLength === 0) {
        app.showDeviceListScreen();
        app.setStatus("No Bluetooth Peripherals Discovered.");
    }
},
```

The connect function is called when the user taps an item in the device list. The device id is retrieved from the element's dataset. The progress indicator is shown, and then the Bluetooth Low Energy API is used to connect to the device. The success callback, `app.connect`, is called when the connection is successful. The failure callback, `app.disconnect`, is called if the connection fails. The failure callback is a long-running callback and will be called anytime the peripheral unexpectedly disconnects.

```
connect: function (e) {
    var device = e.target.dataset.deviceId;
    app.showProgressIndicator("Requesting connection to " + device);
    ble.connect(device, app.onConnect, app.onDisconnect);
},
```

When the application successfully connects to a peripheral, `onConnect` is called. This function saves a reference to the connected peripheral in the app object so that other functions such as `unlock` have access to the peripheral.

The `ble.startNotification` API call is used to subscribe to notifications when the message characteristic changes. Whenever the message characteristic changes, `app.onData` will be called.

```
onConnect: function(peripheral) {
    app.connectedPeripheral = peripheral;
    app.showUnlockScreen();
    app.setStatus("Connected");
    ble.startNotification(peripheral.id, SERVICE_UUID, MESSAGE_UUID, app.onData);
},
```

The `onDisconnect` function is the failure callback from `ble.connect` that is called when the application unexpectedly disconnects.

```
onDisconnect: function(reason) {
    if (!reason) {
        reason = "Connection Lost";
    }
    app.hideProgressIndicator();
    app.showDeviceListScreen();
    app.setStatus(reason);
},
```

The `disconnect` function is called when the user wants to disconnect; it handles the event for the disconnect button. The function sets a status message and begins scanning again after disconnecting from the peripheral.

```
disconnect: function (e) {
    if (e) {
        e.preventDefault();
    }

    app.setStatus("Disconnecting...");
    ble.disconnect(app.connectedPeripheral.id, function() {
        app.setStatus("Disconnected");
        setTimeout(app.scan, 800);
    });
},
```

The `onData` function is the callback that is called whenever the status message characteristic changes. This is how the lock sends status data to the application. The function receives an `ArrayBuffer` and uses the `bytesToString` helper method to convert this to a string. The status message is displayed for the user and the progress indicator is hidden, if it is visible.

```
onData: function(buffer) {
    var message = bytesToString(buffer);
    app.setStatus(message);
    app.hideProgressIndicator();
},
```

The `unlock` function is a bit more complex. Unlock is called when the user submits the form with an unlock code. The unlock code is retrieved from the form data and stored in a variable. The function short-circuits if the unlock code is empty. Success and failure functions are created for the BLE API call.

The API call, `ble.write`, is used to send data to the peripheral. The application writes the unlock code to the unlock characteristic. The write function needs the peripheral id, service UUID, characteristic UUID, and data. The data must be an ArrayBuffer, so the helper function `stringToArrayBuffer` is used to convert the data. The success and failure callback are called based on the results of the write. If the write is successful, the form is simply cleared. The user will hear the lock open and see the LED turn green. The application will receive and display the "unlocked" status from the message characteristic.

```
unlock: function(e) {
    var code = e.target.code.value;
    e.preventDefault(); // don't submit the form

    if (code === "") { return; } // don't send empty data
    app.showProgressIndicator();

    function success() {
        e.target.code.value = ""; //  clear the input
    }

    function failure (reason) {
        alert("Error sending code " + reason);
        app.hideProgressIndicator();
    }

    ble.write(
        app.connectedPeripheral.id,
        SERVICE_UUID,
        UNLOCK_UUID,
        stringToArrayBuffer(code),
        success, failure
    );

},
```

The remaining functions in app deal with the user interface. `showProgressIndicator` takes an optional message and overlays the progress scrim on top of the user interface.

`scrim.hidden=false` ensures the element is displayed. Rules in the CSS ensure the div overlays the screen and is slightly transparent.

```
showProgressIndicator: function(message) {
    if (!message) { message = "Processing"; }
    scrim.firstElementChild.innerHTML = message;
    scrim.hidden = false;
    statusDiv.innerHTML = "";
},
```

`hideProgressIndiator` simply hides the scrim div. It can be safely called even if the progress indicator is not visible.

```
hideProgressIndicator: function() {
    scrim.hidden = true;
},
```

`showDeviceListScreen` ensures the device list screen is visible and the unlock screen is hidden.

```
showDeviceListScreen: function() {
    unlockScreen.hidden = true;
    deviceListScreen.hidden = false;
    app.hideProgressIndicator();
    statusDiv.innerHTML = "";
},
```

`showUnlockScreen` ensures the unlock screen is visible and the device list screen is hidden.

```
showUnlockScreen: function() {
    unlockScreen.hidden = false;
    deviceListScreen.hidden = true;
    app.hideProgressIndicator();
    statusDiv.innerHTML = "";
},
```

`setStatus` logs the status message to the JavaScript console and displays it to the user.

```
setStatus: function(message){
    console.log(message);
    statusDiv.innerHTML = message;
}
}; // end of app
```

The last line of the JavaScript file should call `app.initialize`.

```
app.initialize();
```

See Example 4-5 for the complete listing of *index.js*.

Example 4-5 *Contents of the index.js file*

```
var SERVICE_UUID = 'D270';
var UNLOCK_UUID  = 'D271';
var MESSAGE_UUID = 'D272';
```

```
function stringToArrayBuffer(str) {
    // assuming 8 bit bytes
    var ret = new Uint8Array(str.length);
    for (var i = 0; i < str.length; i++) {
        ret[i] = str.charCodeAt(i);
        console.log(ret[i]);
    }
    return ret.buffer;
}

function bytesToString(buffer) {
    return String.fromCharCode.apply(null, new Uint8Array(buffer));
}

var app = {
    initialize: function() {
        this.bindEvents();
        deviceListScreen.hidden = true;
        unlockScreen.hidden = true;
    },
    bindEvents: function() {
        document.addEventListener('deviceready', this.onDeviceReady, false);
        document.forms[0].addEventListener('submit', this.unlock, false);
    },
    onDeviceReady: function() {
        deviceList.ontouchstart = app.connect; // assume not scrolling
        refreshButton.ontouchstart = app.scan;
        disconnectButton.onclick = app.disconnect;

        app.scan();
    },
    scan: function(e) {
        deviceList.innerHTML = ""; // clear the list
        app.showProgressIndicator("Scanning for Bluetooth Devices...");

        ble.startScan([SERVICE_UUID],
            app.onDeviceDiscovered,
            function() { alert("Listing Bluetooth Devices Failed"); }
        );

        // stop scan after 5 seconds
        setTimeout(ble.stopScan, 5000, app.onScanComplete);

    },
    onDeviceDiscovered: function(device) {
        var listItem, rssi;

        app.showDeviceListScreen();

        console.log(JSON.stringify(device));
        listItem = document.createElement('li');
        listItem.dataset.deviceId = device.id;
        if (device.rssi) {
```

```
            rssi = "RSSI: " + device.rssi + "<br/>";
        } else {
            rssi = "";
        }
        listItem.innerHTML = device.name + "<br/>" + rssi + device.id;
        deviceList.appendChild(listItem);

        var deviceListLength = deviceList.getElementsByTagName('li').length;
        app.setStatus("Found " + deviceListLength +
                    " device" + (deviceListLength === 1 ? "." : "s."));
    },
    onScanComplete: function() {
        var deviceListLength = deviceList.getElementsByTagName('li').length;
        if (deviceListLength === 0) {
            app.showDeviceListScreen();
            app.setStatus("No Bluetooth Peripherals Discovered.");
        }
    },
    connect: function (e) {
        var device = e.target.dataset.deviceId;
        app.showProgressIndicator("Requesting connection to " + device);
        ble.connect(device, app.onConnect, app.onDisconnect);
    },
    onConnect: function(peripheral) {
        app.connectedPeripheral = peripheral;
        app.showUnlockScreen();
        app.setStatus("Connected");
        ble.notify(peripheral.id, SERVICE_UUID, MESSAGE_UUID, app.onData);
    },
    onDisconnect: function(reason) {
        if (!reason) {
            reason = "Connection Lost";
        }
        app.hideProgressIndicator();
        app.showDeviceListScreen();
        app.setStatus(reason);
    },
    disconnect: function (e) {
        if (e) {
            e.preventDefault();
        }

        app.setStatus("Disconnecting...");
        ble.disconnect(app.connectedPeripheral.id, function() {
            app.setStatus("Disconnected");
            setTimeout(app.scan, 800);
        });
    },
    onData: function(buffer) {
        var message = bytesToString(buffer);
        app.setStatus(message);
        app.hideProgressIndicator();
    },
    unlock: function(e) {
```

```
        var code = e.target.code.value;
        e.preventDefault(); // don't submit the form

        if (code === "") { return; } // don't send empty data
        app.showProgressIndicator();

        function success() {
            e.target.code.value = ""; //  clear the input
        }

        function failure (reason) {
            alert("Error sending code " + reason);
            app.hideProgressIndicator();
        }

        ble.write(
            app.connectedPeripheral.id,
            SERVICE_UUID,
            UNLOCK_UUID,
            stringToArrayBuffer(code),
            success, failure
        );

    },
    showProgressIndicator: function(message) {
        if (!message) { message = "Processing"; }
        scrim.firstElementChild.innerHTML = message;
        scrim.hidden = false;
        statusDiv.innerHTML = "";
    },
    hideProgressIndicator: function() {
        scrim.hidden = true;
    },
    showDeviceListScreen: function() {
        unlockScreen.hidden = true;
        deviceListScreen.hidden = false;
        app.hideProgressIndicator();
        statusDiv.innerHTML = "";
    },
    showUnlockScreen: function() {
        unlockScreen.hidden = false;
        deviceListScreen.hidden = true;
        app.hideProgressIndicator();
        statusDiv.innerHTML = "";
    },
    setStatus: function(message){
        console.log(message);
        statusDiv.innerHTML = message;
    }

};

app.initialize();
```

Run the App

Save all the files, and you're ready to run the application. Go back to your terminal or command prompt and start the server by typing phonegap serve from the project directory.

```
xvi:ble-lock don(master)$ phonegap serve
[phonegap] starting app server...
[phonegap] listening on 10.0.1.14:3000
[phonegap]
[phonegap] ctrl-c to stop the server
[phonegap]
```

1. On your phone or tablet, start the PhoneGap Developer App. Verify that the server address listed in the app matches the address from the phonegap server command.

2. Press Connect. You should see a list of devices offering the BLE Lock Service.

3. Click on your device. You should see the unlock page (Figure 4-6).

4. Enter the **12345** and press the unlock button.

5. Click the disconnect button to go back to the device list.

Figure 4-6 *Left: PhoneGap developer app. Center: Device list screen. Right: Unlock screen*

The source code for this chapter is available on GitHub (*https://github.com/MakeBluetooth/ble-lock*).

 If you are interested in an earlier version of this project that uses Serial Port Profile, check out an older version of this project on the Make: Magazine Blog (http://bit.ly/1SaYPrd).

Improving the Lock

This project provides a basic lock example. There are many improvements you could add. From the usability perspective, you could make the mobile application remember the lock connection so the user wouldn't need to select the Bluetooth device every time.

Another possibility could be to automatically unlock when the phone is in close proximity to the lock. You could do this after you learn about beacons in Chapter 10.

The lock has a short numeric passcode that is susceptible to brute-force cracking. You could create a longer passcode that is sent over the wire and use a short PIN to unlock the passcode on the phone. Twenty bytes of data will fit in a Bluetooth characteristic. Other alternatives could be adding a delay between unlock attempts, potentially increasing the delay every time an invalid passcode is entered.

A hardcoded password isn't great. An alternate design could allow the initial passcode to be set when a button was pressed on the hardware. This would require additional software on the lock to enter a setup or configuration mode and additional characteristics for setting the password.

Use this project as a starting point and add enhancements to make your lock even better.

Bleno Lock 5

Bleno (*https://github.com/sandeepmistry/bleno*) is a library for creating Bluetooth Low Energy peripherals with Node.js. This chapter recreates the Bluetooth enabled lock from Chapter 4 using a Raspberry Pi and bleno. In addition to being a small Linux computer, the Raspberry Pi has GPIO (general-purpose input/output) pins that are used to control hardware, similar to an Arduino.

The Lock Service defined in Table 5-1 is the same Lock Service as Table 4-1 in Chapter 4.

Table 5-1 *Lock Service D270*

Characteristic	UUID	Properties
unlock	D271	write
message	D272	notify

The Lock Service does two things. The unlock characteristic allows the lock to be opened by writing a secret code to the characteristic. The status message characteristic allows the lock to send information such as invalid code or unlocked back to the client. Bluetooth clients that subscribe to notifications on the message characteristic are automatically notified whenever a new status message is available.

Hardware

You will need the following hardware to build the lock:

- Raspberry Pi 2 Model B
- USB Bluetooth 4.0 dongle
- Solenoid lock (*http://adafru.it/1512*)

- 12 V, 1A power supply
- Female 2.1mm power adapter (*https://www.adafruit.com/products/368*)
- 5 V, 1A power supply
- Darlington NPN Transistor (TIP120)
- Red LED
- Green LED
- Two 220 Ohm resistors
- Female/male jumper wires (*https://www.adafruit.com/products/826*)

The lock for this project has a solenoid actuated slug. The lock is normally in the closed or locked position. When power is applied to the solenoid, the bolt is drawn back with an electromagnet opening the lock. Wire the lock based on the diagrams in Figures 5-1 and 5-2. Use female-to-male jumper wires to go from the Raspberry Pi headers to the breadboard.

The solenoid lock requires 650mA at 12V to open. The Raspberry Pi takes 5V input and runs at 3.3V. The project has two power supplies. Use a 5V USB micro adapter for the Raspberry Pi. The 12V power supply will plug into the 2.1mm jack on the female DC power adapter. Run the 12V positive wire from the screw terminal on the adapter directly to the solenoid. Run the 12V negative to the ground rail on the breadboard. The 5V and 12V system share a common ground connection. The Raspberry Pi uses a 3.3V pin to switch a Darlington transistor, sending the higher voltage and current from the 12V power supply to the solenoid.

Figure 5-1 *Lock wiring for Raspberry Pi*

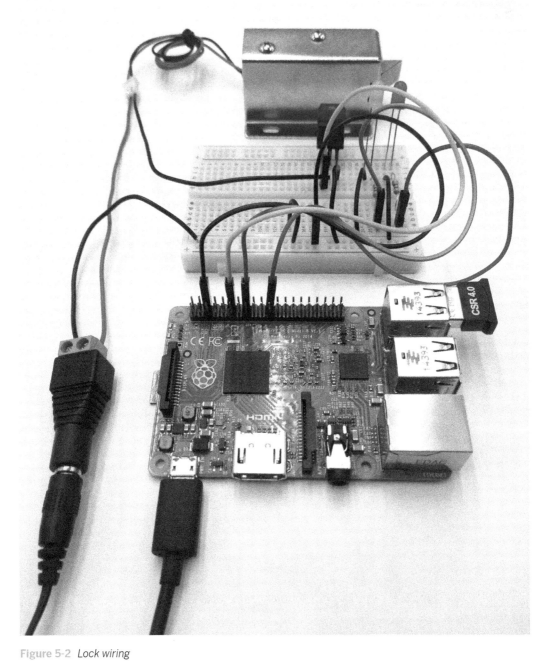

Figure 5-2 *Lock wiring*

Red and green LEDs are used to show the lock status. The green LED will light when the unlock code is correct. The red LED will light when the code is incorrect. Both the LEDs and the lock reset after four seconds.

 This project uses the Raspberry Pi 2 Model B. If you have a different Raspberry Pi, it should work fine; just double-check the GPIO pin numbers when wiring. This code should also be able to run on a BeagleBone Black with minimal changes.

The Raspberry Pi doesn't come with Bluetooth, but it's easy to add with a USB adapter. Most any Linux-compatible 4.0 Bluetooth adapter should work. I use a generic bluetooth 4.0 adapter (*https://www.adafruit.com/products/1327*) from Adafruit that has a CSR8510 chip (Figure 5-3).

Figure 5-3 *Generic Bluetooth adapter*

Lock Software

Ensure that Node.js and the prerequisities for bleno are installed on your Raspberry Pi. Details are covered in "Installing Node.js".

Libraries

Open a terminal on your Raspberry Pi. Create a new directory named lock for the project. Change into the lock directory and use npm to install bleno and onoff (*https://www.npmjs.com/package/onoff*) into the project. The onoff package is used for GPIO access.

```
pi@raspberrypi ~ $ mkdir lock
pi@raspberrypi ~ $ cd lock
pi@raspberrypi ~/lock $ npm install bleno
pi@raspberrypi ~/lock $ npm install onoff
```

Programming

Create a new file named *lock.js* and open it in your text editor. Require the util and bleno packages at the beginning of the file.

```
var util = require('util');
var bleno = require('bleno');
```

Create an instance of Gpio and define the pins for the LEDs and the lock. These numbers correspond to the pins in Figure 5-1.

```
var Gpio = require('onoff').Gpio,
    greenLed = new Gpio(23, 'out'),
    redLed = new Gpio(25, 'out'),
    lock = new Gpio(18, 'out');
```

Create vars to alias bleno's PrimaryService, Characteristic, and Descriptor objects.

```
var PrimaryService = bleno.PrimaryService;
var Characteristic = bleno.Characteristic;
var Descriptor = bleno.Descriptor;
```

The software hardcodes the passcode used to open the lock.

```
var secret = '12345';
```

Create the unlock characteristic. The UUID and information comes from the first row of the service definition (Table 5-1). The descriptor is not required but helps make the characteristic self-documenting for Bluetooth clients. If you're not a JavaScript developer, some of the syntax might not be familiar. util.inherits allows the UnlockCharacteristic to inherit from Characteristic, including the constructor and other methods.

```
var UnlockCharacteristic = function() {
    UnlockCharacteristic.super_.call(this, {
      uuid: 'd271',
      properties: ['write'],
      descriptors: [
```

```
        new Descriptor({
          uuid: '2901',
          value: 'Unlock'
        })
      ]
    });
  };
  util.inherits(UnlockCharacteristic, Characteristic);
```

Create the `onWriteRequest` function. Bleno calls `onWriteRequest` when a client writes a new value to the unlock characteristic.

This function checks the received data against the secret code. If the passcode matches, the green LED is lit and the lock is opened. Otherwise, the red LED is turned on. A timer is set to reset the lock state after four seconds. A success result is sent to the bleno callback. The status characteristic will handle the status event that is fired.

```
UnlockCharacteristic.prototype.onWriteRequest = function(data, offset, withoutResponse,
callback) {
  var status;

  if (data.toString() === secret) {
    status = 'unlocked';
    greenLed.writeSync(1);
    lock.writeSync(1);
  } else {
    status = 'invalid code';
    redLed.writeSync(1);
  }

  // reset lock and lights after 4 seconds
  setTimeout(this.reset.bind(this), 4000);

  console.log('unlock: ' + data);
  console.log('status: ' + status);

  callback(this.RESULT_SUCCESS);

  this.emit('status', status);
};
```

Create the `reset` function. Reset's job is to close the lock, turn off the LEDs, and emit the locked status event.

```
UnlockCharacteristic.prototype.reset = function() {
  this.emit('status', 'locked');
  lock.writeSync(0);
  redLed.writeSync(0);
  greenLed.writeSync(0);
}
```

Create the status characteristic. The data for this characteristic is from the second row of the service definition (Table 5-1). This characteristic has the `notify` property.

Note that the unlockCharacteristic is passed to the function that creates the Status Characteristic. The onUnlockStatusChange function is added as an event listener on the unlock Characteristic. This allows the status characteristic to handle status events from the unlock characteristic.

```
var StatusCharacteristic = function(unlockCharacteristic) {
    StatusCharacteristic.super_.call(this, {
      uuid: 'd272',
      properties: ['notify'],
      descriptors: [
        new Descriptor({
          uuid: '2901',
          value: 'Status Message'
        })
      ]
    });

    unlockCharacteristic.on('status', this.onUnlockStatusChange.bind(this));
  };
util.inherits(StatusCharacteristic, Characteristic);
```

The onUnlockStatusChange function handles status events. When an event is processed, the characteristic's value is set to the status that was passed to the function. Since this characteristic has the notify property, bleno notifies the connected client when the status changes.

```
StatusCharacteristic.prototype.onUnlockStatusChange = function(status) {
  if (this.updateValueCallback) {
    this.updateValueCallback(new Buffer(status));
  }
};
```

All the code up to this point has been to set up the characteristics. Use those functions to create instances of unlock and status characteristics.

```
var unlockCharacteristic = new UnlockCharacteristic();
var statusCharacteristic = new StatusCharacteristic(unlockCharacteristic);
```

Define the Lock Service as a primary service. The Lock Service includes the unlock and status characteristics.

```
var lockService = new PrimaryService({
  uuid: 'd270',
  characteristics: [
    unlockCharacteristic,
    statusCharacteristic
  ]
});
```

Register for bleno's stateChange event. When the application starts, bleno fires the state Change event with the current state of the Bluetooth adapter. If the Bluetooth adapter is powered on, the application starts advertising a peripheral named RPi Lock offering the Lock Service.

```
bleno.on('stateChange', function(state) {
  console.log('on -> stateChange: ' + state);

  if (state === 'poweredOn') {
    bleno.startAdvertising('RPi Lock', [lockService.uuid]);
  } else {
    bleno.stopAdvertising();
  }
});
```

Once bleno starts advertising, it fires the advertisingStart event. As long as there are no errors, the lockService gets added to bleno's services.

```
bleno.on('advertisingStart', function(error) {
  console.log('on -> advertisingStart: ' + (error ? 'error ' + error : 'success'));

  if (!error) {
    bleno.setServices([lockService]);
  }
});
```

The last thing we need to do is some cleanup. The program can be terminated with CTRL +C. When the application exits, it should release the GPIO pins to free resources. The exit function is bound to the SIGINT event, so it automatically runs when CTRL+C is used to terminate the application.

```
function exit() {
  greenLed.unexport();
  redLed.unexport();
  lock.unexport();
  process.exit();
}
process.on('SIGINT', exit);
```

Save your file.

Run *lock.js* on your Raspberry Pi using Node.js. You should see output similar to Figure 5-4.

```
pi@raspberrypi ~/lock $ sudo node lock.js
```

Figure 5-4 *Bleno lock successful start*

Now that the lock is running, connect with a generic Bluetooth client like LightBlue (Figure 5-5) or nRF Master Control Panel. Connect to Lock Service 0xD270. Subscribe to the Status Characteristic 0x272. Write the passcode to the Unlock Characteristic 0xD271. If you send *12345* as text, the lock should open. See "Testing the Lock" in Chapter 4 for more details.

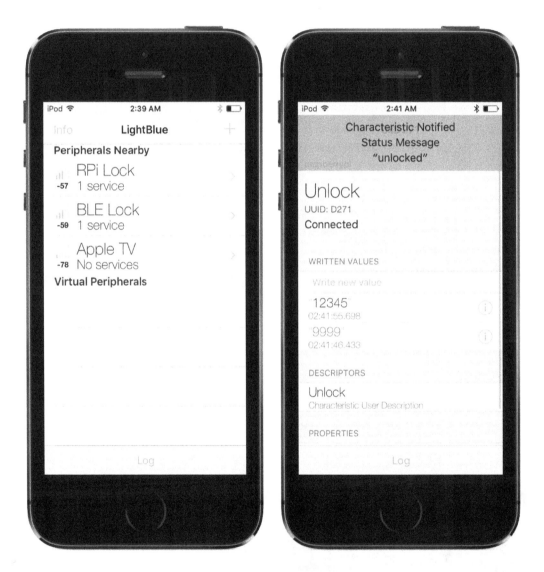

Figure 5-5 *Left: Discovering locks with LightBlue. Right: Unlocking with LightBlue.*

If you built the PhoneGap lock app ("BLE Lock App") from Chapter 4, it will unlock this lock too. The PhoneGap app connects to the Lock Service; it doesn't care if the lock is implemented on Arduino or Raspberry Pi (Figure 5-6).

Figure 5-6 *Left: Discovering locks with PhoneGap. Right: Unlocking with PhoneGap.*

Once you connect and start sending codes to the lock, you should see output in the terminal on the Raspberry Pi (Figure 5-7).

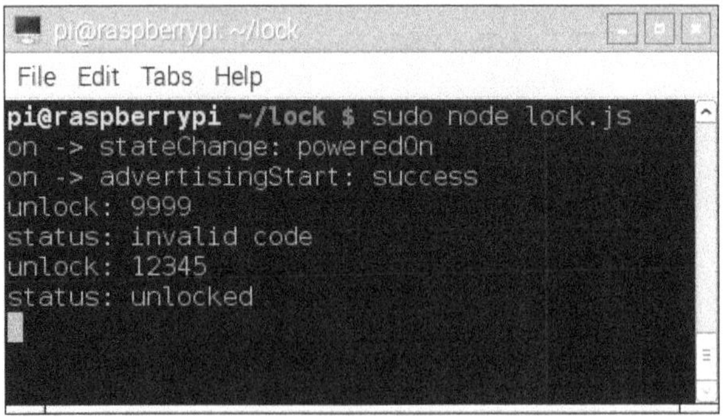

Figure 5-7 *Bleno lock with output from connected client*

The source code for this chapter is available on GitHub (*https://github.com/MakeBluetooth/ bleno-lock*).

Conclusion

This project replicated the Bluetooth lock from Chapter 4. Since both versions of the lock implement the same Bluetooth Lock Service (see Table 5-1), the new lock works with the BLE Lock PhoneGap app. Bluetooth clients access the service; the implementation of the service doesn't matter.

The bleno implementation of the Lock Service was longer and a bit more complex than the Arduino version. The Arduino version runs on a dedicated microcontroller. The Raspberry Pi is a (relatively) powerful networked computer with plenty of memory, lots of storage, and a full Linux installation. Each implementation is good—they're just different. Your application requirements can help determine which hardware to choose for your next Bluetooth project.

Weather Station §

In this chapter, you will build a Bluetooth-LE-enabled weather station. A BME280 sensor (*https://www.adafruit.com/products/2652*) will be used to measure temperature, humidity, and pressure. The Adafruit BME280 breakout board combines a Bosch BME280 environmental sensor with a voltage regulator and other components to make it easier to use.

The Weather Station Service is defined in Table 6-1. Bluetooth Low Energy clients can read the values of the sensors at any time. When the client first connects, it reads the characteristic values to get the initial data. The client could periodically re-read the characteristics to see if any data changed, but it is not very efficient for the client to poll the server looking for changes in values. Each sensor characteristic has the `notify` property set. The client can subscribe to notifications and the Arduino weather station will notify the connected client whenever a characteristic value changes.

Our weather station measures temperature, humidity, and pressure. More sophisticated weather stations measure wind speed, wind direction, rainfall, and dew point. If you'd like, you can expand your weather station with additional sensors. Additional sensors will require new characteristics in the Bluetooth service. This sketch maxes out the available memory on the Arduino, so a more sophisticated weather station may require a different Arduino board with more memory.

Table 6-1 *Weather Station Service BBB0*

Characteristic	UUID	Properties	Comment
Temperature	BBB1	read, notify	Temperature (degrees C)
Humidity	BBB2	read, notify	Percent relative humidity
Pressure	BBB3	read, notify	Barometric pressure (pascals)

Hardware

The hardware for this project starts with the Arduino Uno and Adafruit Bluefruit LE nRF8001 radio we have used for other projects. Refer to "Wiring Up the Adafruit Bluefruit LE Module" for more information. Wire in the BME280 breakout board sensor based on Figures 6-1 and 6-2.

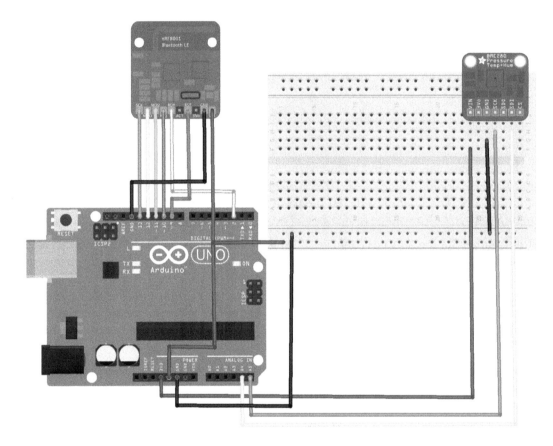

Figure 6-1 *Weather station wiring diagram*

Figure 6-2 *Weather station wiring*

The BME280 can use SPI or I2C. This project uses I2C. The pin labeled SCK is the I2C clock pin and should be connected to A4 or the SCL pin on the Uno. The pin labeled SDI is a I2C data pin and should be connected to A5 or the SDA pin on the Uno.

Libraries

If you've done another Arduino project in this book, you should have the Arduino BLE Peripheral library installed. If not, refer to "Installing the BLE Peripheral Library" for details on how to install the library.

Install the Adafruit BME280 library (Figure 6-3) and the Adafruit Universal Sensor library (Figure 6-4) using the Arduino Library Manager. To get there, go to Sketch → Include Library → Manage Libraries....

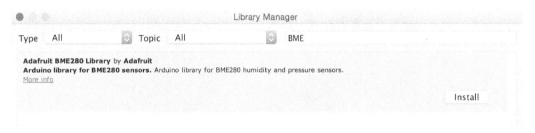

Figure 6-3 *Library Manager for the BME280 Sensor library*

Figure 6-4 *Library Manager for the Unified Sensor library*

Programming

Now that the hardware is in place, you need to write the code. Open the Arduino IDE and create a new sketch using File → New. Save the sketch as "Weather."

At the beginning of the sketch, before setup, include the Bluetooth library, define the pins, and create a BLE peripheral.

```
#include <SPI.h>
#include <BLEPeripheral.h>

#define BLE_REQ 10
#define BLE_RDY 2
#define BLE_RST 9

BLEPeripheral blePeripheral = BLEPeripheral(BLE_REQ, BLE_RDY, BLE_RST);
```

```
    if (!isnan(pressure) && pressureCharacteristic.value() != pressure) {
      pressureCharacteristic.setValue(pressure);
      Serial.print(F("Pressure "));
      Serial.println(pressure);
    }

  }
```

See Example 6-1 for the complete listing of the weather sketch.

Example 6-1 *Weather.ino*

```
#include <SPI.h>
#include <BLEPeripheral.h>

// define pins for Adafruit Bluefruit LE
// https://github.com/sandeepmistry/arduino-BLEPeripheral#pinouts
#define BLE_REQ 10
#define BLE_RDY 2
#define BLE_RST 9

BLEPeripheral blePeripheral = BLEPeripheral(BLE_REQ, BLE_RDY, BLE_RST);
BLEService weatherService = BLEService("BBB0");
BLEFloatCharacteristic temperatureCharacteristic =
    BLEFloatCharacteristic("BBB1", BLERead | BLENotify);
BLEDescriptor temperatureDescriptor = BLEDescriptor("2901", "Temp");
BLEFloatCharacteristic humidityCharacteristic =
    BLEFloatCharacteristic("BBB2", BLERead | BLENotify);
BLEDescriptor humidityDescriptor = BLEDescriptor("2901", "Humidity");
BLEFloatCharacteristic pressureCharacteristic =
    BLEFloatCharacteristic("BBB3", BLERead | BLENotify);
BLEDescriptor pressureDescriptor = BLEDescriptor("2901", "Pressure");

#include <Wire.h>
#include <Adafruit_Sensor.h>
#include <Adafruit_BME280.h>
Adafruit_BME280 bme;

long previousMillis = 0;  // stores the last time sensor was read
long interval = 2000;     // interval at which to read sensor (milliseconds)

void setup()
{
  Serial.begin(9600);
  Serial.println(F("Bluetooth Low Energy Weather Station"));

  // set advertised name and service
  blePeripheral.setLocalName("Weather");
  blePeripheral.setDeviceName("Weather");
  blePeripheral.setAdvertisedServiceUuid(weatherService.uuid());

  // add service and characteristic
  blePeripheral.addAttribute(weatherService);
```

```
blePeripheral.addAttribute(temperatureCharacteristic);
blePeripheral.addAttribute(temperatureDescriptor);
blePeripheral.addAttribute(humidityCharacteristic);
blePeripheral.addAttribute(humidityDescriptor);
blePeripheral.addAttribute(pressureCharacteristic);
blePeripheral.addAttribute(pressureDescriptor);

blePeripheral.begin();
if (!bme.begin()) {
  Serial.println(F("Could not find a valid BME280 sensor, check wiring!"));
  while (1);
}
}

void loop()
{
  // Tell the bluetooth radio to do whatever it should be working on
  blePeripheral.poll();

  // limit how often we read the sensor
  if (millis() - previousMillis > interval) {
    pollSensors();
    previousMillis = millis();
  }
}

void pollSensors()
{

  float temperature = bme.readTemperature();
  float humidity = bme.readHumidity();
  float pressure = bme.readPressure();

  // only set the characteristic value if the temperature has changed
  if (!isnan(temperature) && temperatureCharacteristic.value() != temperature) {
    temperatureCharacteristic.setValue(temperature);
    Serial.print(F("Temperature "));
    Serial.println(temperature);
  }

  // only set the characteristic value if the humidity has changed
  if (!isnan(humidity) && humidityCharacteristic.value() != humidity) {
    humidityCharacteristic.setValue(humidity);
    Serial.print(F("Humidity "));
    Serial.println(humidity);
  }

  // only set the characteristic value if the pressure has changed
  if (!isnan(pressure) && pressureCharacteristic.value() != pressure) {
    pressureCharacteristic.setValue(pressure);
    Serial.print(F("Pressure "));
    Serial.println(pressure);
  }
```

```
    }
```

Compile and Upload

Compile and upload the code to your Arduino via File → Upload.

The source code for this chapter is available on GitHub (*https://github.com/MakeBluetooth/ble-weather*).

Serial Monitor

While the sketch is running, open the Arduino serial monitor to view the temperature, humidity, and pressure as the values change (see Figure 6-5).

```
Tools -> Serial Monitor
```

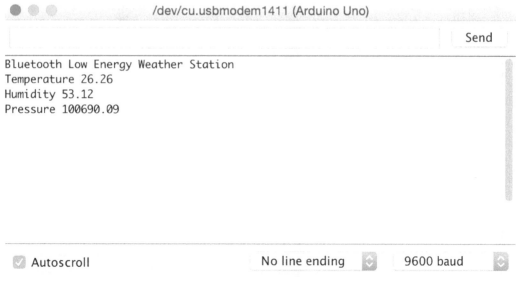

Figure 6-5 *Arduino IDE serial monitor*

Using the Service

The serial monitor shows that data is read from the sensor and processed with your sketch. The next step is to use a generic Bluetooth client to connect to the service and read data from the Bluetooth characteristics. If you have an iOS device, continue to the next section on LightBlue. Android users should skip ahead to "Using nRF Master Control Panel on Android".

Using LightBlue on iOS

When you first connect with LightBlue, it will be difficult to interpret the data because the UI displays hex bytes for the temperature, humidity, and pressure value. Select the temperature characteristic and tap on the Hex button on the top-right. Scroll down and select 4 Byte Float Little Endian. The UI will switch back to the characteristic view and display the temperature as a float (Figure 6-6).

Figure 6-6   *The temperature displays as a floating-point number*

Using nRF Master Control Panel on Android

On Android, we read the data with nRF Master Control Panel (Figure 6-7).

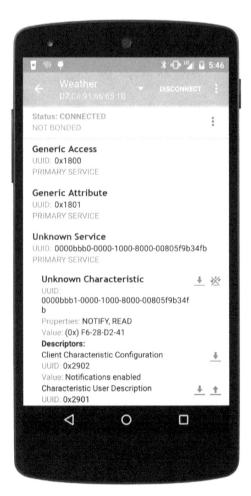

Figure 6-7 *Click the icon with the three down arrows to enable temperature notifications*

Unfortunately, nRF Master Control Panel doesn't translate the bytes of data into floating-point numbers. You can use the IEEE 754 Float Converter (*http://bit.ly/1NHmTxW*) to convert the hexadecimal characteristic value to a float. Type the value in the Hexadecimal Representation input box (Figure 6-8).

Figure 6-8 *Float Converter*

Note that you need to reverse the byte order between nRF Master Control Panel and the Float Converter. nRF Master Control Panel displays the data in network byte order (big-endian). The Float Converter expects the most significant byte first (little-endian).

nRF Master Control Panel	(0x)F6-28-D2-41
Hexadecimal Representation	0x41d228f6
Temperature	26.3 °C

PhoneGap

Generic Bluetooth applications, like LightBlue, can read data from the weather station, but the user interface makes reading multiple values tedious. You can create a PhoneGap application that will run on iOS or Android and display the weather station values.

Refer to "Installing PhoneGap" for more information on installing the PhoneGap command-line tools on your computer.

Create the Project

Open a terminal or cmd prompt on your computer. Change to the directory where you'd like to create the project.

```
cd ~/bluetooth
```

Use the phonegap command-line tool to create a new project.

```
phonegap create weather com.makebluetooth.weather Weather --template blank
```

Since you are using the PhoneGap Developer App (*http://app.phonegap.com*), there's no need to install the Bluetooth Low Energy plugin or any platform-specific code. See "Installing PhoneGap" for more details about using Apache Cordova or PhoneGap to build native applications.

Open the project in your text editor.

HTML

The *index.html* page builds the user interface for the mobile application. The application has two pages. The first page lists the available devices that are running the 0xBBB0 Bluetooth Weather Service. When you tap on a device in the list, you are connected to the device and taken to the detail page. The detail page shows data from the Bluetooth weather service.

Application pages are created using HTML `div` elements. The elements get unique ids so it is easy to access them from the JavaScript code. JavaScript functions show and hide divs as the user navigates in the application.

Add a link to the stylesheet in the head section of *index.html*.

```
<link rel="stylesheet" type="text/css" href="css/index.css" />
```

The stylesheet isn't strictly required, but the application will look better if you add it. Create a new *css* directory inside the *www* directory. Create a file named *index.css*. Copy the contents of *index.css* (*http://bit.ly/1SaZARe*) into your file. This book already covers a lot of topics. Copying in some premade CSS allows us to skip the explanation and get back to building the application.

Edit the code within the body tag of *index.html*.

An `h2` header element serves as a title bar for the application.

```
<h2>BLE Weather</h2>
```

The main page contains a list for devices and a refresh button.

```
<div id="mainPage">
    <ul id="deviceList">
    </ul>
    <button id="refreshButton">Refresh</button>
</div>
```

The detail page is slightly more complex. It contains a series of nested `div` elements to display the data. JavaScript functions will fill in this data as the application runs.

```
<div id="detailPage">
    <p>Temperature</p>
    <div id="temperatureDiv">
        Unknown
    </div>
    <p>Humidity</p>
    <div id="humidityDiv">
        Unknown
    </div>
    <p>Station Pressure</p>
    <div id="stationPressureDiv">
        Unknown
    </div>
    <p>Sea Level Pressure</p>
    <div id="seaLevelPressureDiv">
```

```
        Unknown
    </div>
    <button id="disconnectButton">Disconnect</button>
</div>
```

Finally, ensure that both *cordova.js* and *index.js* are included just before the close body tag.

```
<script type="text/javascript" src="cordova.js"></script>
<script type="text/javascript" src="js/index.js"></script>
```

See Example 6-2 for the complete listing of *index.html*.

Example 6-2 *index.html*

```
<!DOCTYPE html>
<html>
    <head>
        <meta charset="utf-8">
        <meta name="viewport" content="initial-scale=1, maximum-scale=1,
         user-scalable=no, width=device-width">
        <link rel="stylesheet" type="text/css" href="css/index.css" />
        <title>BLE Weather</title>
    </head>
    <body>
        <h2>BLE Weather</h2>
        <div id="mainPage">
            <ul id="deviceList">
            </ul>
            <button id="refreshButton">Refresh</button>
        </div>
        <div id="detailPage">
            <p>Temperature</p>
            <div id="temperatureDiv">
                Unknown
            </div>
            <p>Humidity</p>
            <div id="humidityDiv">
                Unknown
            </div>
            <p>Station Pressure</p>
            <div id="stationPressureDiv">
                Unknown
            </div>
            <p>Sea Level Pressure</p>
            <div id="seaLevelPressureDiv">
                Unknown
            </div>
            <button id="disconnectButton">Disconnect</button>
        </div>
        <script type="text/javascript" src="cordova.js"></script>
        <script type="text/javascript" src="js/index.js"></script>
```

```
      </body>
   </html>
```

JavaScript

PhoneGap applications are written in JavaScript. Create a new *js* directory inside the *www* directory. Create a new file named *index.js*. Define four constants for the Bluetooth service and characteristics at the beginning of the file.

```
var WEATHER_SERVICE = 'BBB0';
var TEMPERATURE_CHARACTERISTIC = 'BBB1';
var HUMIDITY_CHARACTERISTIC = 'BBB2';
var PRESSURE_CHARACTERISTIC = 'BBB3';
```

Create an object literal named app. The application logic is built by adding functions to the app object.

```
var app = {
};
```

The first three functions you add are common bootstrapping for most PhoneGap apps. The initialize function calls bindEvents and ensures the main page is visible. The bindE vents function adds JavaScript event listeners to the list and buttons. The onDeviceReady function is called when the deviceready event fires. PhoneGap applications must wait for the framework to initialize and fire deviceready before making any API calls.

```
initialize: function() {
    this.bindEvents();
    this.showMainPage();
},
bindEvents: function() {
    document.addEventListener('deviceready', this.onDeviceReady, false);
    deviceList.addEventListener('click', this.connect, false);
    refreshButton.addEventListener('click', this.refreshDeviceList, false);
    disconnectButton.addEventListener('click', this.disconnect, false);
},
onDeviceReady: function() {
    app.refreshDeviceList();
},
```

Once the device is ready, the application can scan for Bluetooth devices and display the results in the user interface. Bluetooth scan results are filtered to contain only devices that are advertising the 0xBBB0 Weather Service.

```
refreshDeviceList: function() {
    deviceList.innerHTML = ''; // empty the list
    ble.scan([WEATHER_SERVICE], 5, app.onDiscoverDevice, app.onError);
},
```

On Android 4.3 and 4.4, scan filtering is broken. Often you cannot filter scan results by UUID. If you are not seeing any devices while scanning, try passing an empty array [] to ble.scan *instead of filtering for* [WEATH ER_SERVICE].

The onDiscoverDevice callback will be called for each Bluetooth device that is found. Add a li element to the unordered list that will display the device name and id. The device id is stored in the list item's dataset so that it can be retrieved later.

```
onDiscoverDevice: function(device) {
    var listItem = document.createElement('li');
    listItem.innerHTML = device.name + ' ' + device.id;
    listItem.dataset.deviceId = device.id;
    deviceList.appendChild(listItem);
},
```

When a user taps an item in the list, the connect function is called. The device id is retrieved from the dataset and passed to the connect function of the ble API.

```
connect: function(e) {
    var deviceId = e.target.dataset.deviceId;
    ble.connect(deviceId, app.onConnect, app.onError);
},
```

When the application connects to the Bluetooth peripheral, the success callback app.on Connect is called. After a successful connection, a few things need to happen. First, save a reference to the connected peripheral in app.peripheral. You'll need the reference to the peripheral later in the disconnect method. Next, display the detail page to the user using app.showDetailPage (you'll write this function later.) Use ble.startNotification to subscribe to notifications when the temperature, humidity, or pressure changes. It's also important to call ble.read for each characteristic. If the values are not changing, ble.read ensures there is data to show the user. The read and notify for each property can share the same success callback.

```
onConnect: function(peripheral) {
    app.peripheral = peripheral;
    app.showDetailPage();

    var failure = function(reason) {
        navigator.notification.alert(reason, null, "Temperature Error");
    };

    // subscribe to be notified when the temperature changes
    ble.startNotification(
        peripheral.id,
        WEATHER_SERVICE,
        TEMPERATURE_CHARACTERISTIC,
        app.onTemperatureChange,
        failure
```

```
    );

    // subscribe to be notified when the humidity changes
    ble.startNotification(
        peripheral.id,
        WEATHER_SERVICE,
        HUMIDITY_CHARACTERISTIC,
        app.onHumidityChange,
        failure
    );

    // subscribe to be notified when the pressure changes
    ble.startNotification(
        peripheral.id,
        WEATHER_SERVICE,
        PRESSURE_CHARACTERISTIC,
        app.onPressureChange,
        failure
    );

    // read the initial values
    ble.read(
        peripheral.id,
        WEATHER_SERVICE,
        TEMPERATURE_CHARACTERISTIC,
        app.onTemperatureChange,
        failure
    );

    ble.read(
        peripheral.id,
        WEATHER_SERVICE,
        HUMIDITY_CHARACTERISTIC,
        app.onHumidityChange,
        failure
    );

    ble.read(
        peripheral.id,
        WEATHER_SERVICE,
        PRESSURE_CHARACTERISTIC,
        app.onPressureChange,
        failure
    );

},
```

Now, implement the callback for the `ble.read` and `ble.startNotification` calls. All three characteristics are floating-point numbers. Each callback receives an `ArrayBuffer` of data. This buffer of raw data needs to be converted into a typed `Float32Array` so it can be read. Although the ArrayBuffer is converted into an array of floating-point values, the service only sends one value, so it is an array of one element.

Convert temperature into a floating-point value. Temperature is sent in Celsius. Convert this to Fahrenheit and display both values in the application. Use toFixed to truncate the number to one decimal place.

```
onTemperatureChange: function(buffer) {
    var data = new Float32Array(buffer);
    var temperature = data[0];
    var temperatureF = temperature * 1.8 + 32;
    var message = temperature.toFixed(1) + "&deg;C<br/>" +
      temperatureF.toFixed(1) + "&deg;F<br/>";
    temperatureDiv.innerHTML = message;
},
```

Humidity is a floating-point value like temperature. Use toFixed to display one decimal place.

```
onHumidityChange: function(buffer) {
    var data = new Float32Array(buffer);
    var humidity = data[0];
    var message = humidity.toFixed(1) + "%";
    humidityDiv.innerHTML = message;
},
```

The handler for pressure is slightly more complex. The pressure value is just a floating-point number like temperature and humidity, but there are a few more conversions that need to be done.

Data is sent in pascals, but hectoPascals (or millbars) is a better way to display pressure information. Divide by 100 to get hPa. Displaying pressure in hPa works fine for most of the world. Unfortunately, the United States still uses the antiquated unit *inches of mercury* to measure pressure. Use app.toInchesOfMercury helper function to do the conversion.

The pressure measured by the sensor is known as station pressure. This measured value probably does not match the value you look up on a weather website such as NOAA (*http://noaa.gov*), GoC Weather (*http://weather.gc.ca*), or BBC Weather (*http://www.bbc.com/weather*).

Atmospheric pressure decreases as elevation increases. Barometric pressure is reported at sea-level pressure as a way to normalize pressure measurements, removing the effect of the station's elevation. Once you know the elevation of your weather station, you can convert the pressure to theoretical pressure at sea level.

There are formulas for calculating sea-level pressure from station pressure using temperature and elevation. As a rough estimate, assume that pressure will decrease 1 hPa for every 8 meters of elevation. If you are at higher elevation or need a more accurate value, you may need to substitute a better conversion formula.

Find your current elevation using a website like What Is My Elevation? (http://www.whatismyelevation.com)

```
onPressureChange: function(buffer) {
    var data = new Float32Array(buffer);
    var pressure = data[0]; // pascals

    // hectoPascals (or millibar) is a better unit of measure
    pressure = pressure / 100.0;

    // station pressure is what we measure
    var message = pressure.toFixed(2) + " hPa<br/>" +
        app.toInchesOfMercury(pressure) + " inHg<br/>";

    stationPressureDiv.innerHTML = message;

    // Pressure needs to be converted to sea-level pressure to match
    // the barometric pressure in weather reports.

    // Fort Washington, PA - adjust this for your location
    var elevationInMeters = 50.438;

    // pressure drops approx 1 millibar for every 8 meters above sea level
    var delta = elevationInMeters / 8.0;
    var seaLevelPressure = pressure + delta;

    message = seaLevelPressure.toFixed(2) + " hPa<br/>" +
      app.toInchesOfMercury(seaLevelPressure) + " inHg";

    seaLevelPressureDiv.innerHTML = message;
},
```

Create a helper function to covert between hectoPascals and inches of mercury. This can be estimated by using the conversion factor of 0.0295300 inHg per hPa.

```
toInchesOfMercury: function(hPa) {
  // http://www.srh.noaa.gov/images/epz/wxcalc/pressureConversion.pdf
  return (hPa * 0.0295300).toFixed(2);
},
```

Create a function for the disconnect button.

```
disconnect: function(e) {
    if (app.peripheral && app.peripheral.id) {
        ble.disconnect(app.peripheral.id, app.showMainPage, app.onError);
    }
},
```

Create two functions for page navigation. Navigation simply shows and hides divs.

```
showMainPage: function() {
    mainPage.hidden = false;
```

```
        detailPage.hidden = true;
    },
    showDetailPage: function() {
        mainPage.hidden = true;
        detailPage.hidden = false;
    },
```

The last function in the app object is the generic error handler referenced from many of the API calls. If an error occurs, a dialog box is shown to the user.

```
onError: function(reason) {
    navigator.notification.alert(reason, app.showMainPage, "Error");
}
```

The last line of the JavaScript file should initialize the application.

```
app.initialize();
```

See Example 6-3 for the complete listing of *index.js*.

Example 6-3 *Full listing of index.js*

```
var WEATHER_SERVICE = 'BBB0';
var TEMPERATURE_CHARACTERISTIC = 'BBB1';
var HUMIDITY_CHARACTERISTIC = 'BBB2';
var PRESSURE_CHARACTERISTIC = 'BBB3';

var app = {
    initialize: function() {
        this.bindEvents();
        this.showMainPage();
    },
    bindEvents: function() {
        document.addEventListener('deviceready', this.onDeviceReady, false);
        deviceList.addEventListener('click', this.connect, false);
        refreshButton.addEventListener('click', this.refreshDeviceList, false);
        disconnectButton.addEventListener('click', this.disconnect, false);
    },
    onDeviceReady: function() {
        app.refreshDeviceList();
    },
    refreshDeviceList: function() {
        deviceList.innerHTML = ''; // empty the list
        ble.scan([WEATHER_SERVICE], 5, app.onDiscoverDevice, app.onError);
    },
    onDiscoverDevice: function(device) {
        var listItem = document.createElement('li');
        listItem.innerHTML = device.name + ' ' + device.id;
        listItem.dataset.deviceId = device.id;
        deviceList.appendChild(listItem);
    },
    connect: function(e) {
        var deviceId = e.target.dataset.deviceId;
        ble.connect(deviceId, app.onConnect, app.onError);
```

```
    },
onConnect: function(peripheral) {
    app.peripheral = peripheral;
    app.showDetailPage();

    var failure = function(reason) {
        navigator.notification.alert(reason, null, "Temperature Error");
    };

    // subscribe to be notified when the temperature changes
    ble.startNotification(
        peripheral.id,
        WEATHER_SERVICE,
        TEMPERATURE_CHARACTERISTIC,
        app.onTemperatureChange,
        failure
    );

    // subscribe to be notified when the humidity changes
    ble.startNotification(
        peripheral.id,
        WEATHER_SERVICE,
        HUMIDITY_CHARACTERISTIC,
        app.onHumidityChange,
        failure
    );

    // subscribe to be notified when pressure changes
    ble.startNotification(
        peripheral.id,
        WEATHER_SERVICE,
        PRESSURE_CHARACTERISTIC,
        app.onPressureChange,
        failure
    );

    // read the initial values
    ble.read(
        peripheral.id,
        WEATHER_SERVICE,
        TEMPERATURE_CHARACTERISTIC,
        app.onTemperatureChange,
        failure
    );

    ble.read(
        peripheral.id,
        WEATHER_SERVICE,
        HUMIDITY_CHARACTERISTIC,
        app.onHumidityChange,
        failure
    );

    ble.read(
```

```
                peripheral.id,
                WEATHER_SERVICE,
                PRESSURE_CHARACTERISTIC,
                app.onPressureChange,
                failure
            );

    },
    onTemperatureChange: function(buffer) {
        var data = new Float32Array(buffer);
        var temperature = data[0];
        var temperatureF = temperature * 1.8 + 32;
        var message = temperature.toFixed(1) + "&deg;C<br/>" +
            temperatureF.toFixed(1) + "&deg;F<br/>";
        temperatureDiv.innerHTML = message;
    },
    onHumidityChange: function(buffer) {
        var data = new Float32Array(buffer);
        var humidity = data[0];
        var message = humidity.toFixed(1) + "%";
        humidityDiv.innerHTML = message;
    },
    onPressureChange: function(buffer) {
        var data = new Float32Array(buffer);
        var pressure = data[0]; // pascals

        // hectoPascals (or millibar) is a better unit of measure
        pressure = pressure / 100.0;

        // station pressure is what we measure
        var message = pressure.toFixed(2) + " hPa<br/>" +
            app.toInchesOfMercury(pressure) + " inHg<br/>";

        stationPressureDiv.innerHTML = message;

        // Pressure needs to be converted to sea-level pressure to match
        // the barometric pressure in weather reports.

        // Fort Washington, PA - adjust this for your location
        var elevationInMeters = 50.438;

        // pressure drops approx 1 millibar for every 8 meters above sea level
        var delta = elevationInMeters / 8.0;
        var seaLevelPressure = pressure + delta;

        message = seaLevelPressure.toFixed(2) + " hPa<br/>" +
            app.toInchesOfMercury(seaLevelPressure) + " inHg";

        seaLevelPressureDiv.innerHTML = message;
    },
    toInchesOfMercury: function(hPa) {
      // http://www.srh.noaa.gov/images/epz/wxcalc/pressureConversion.pdf
      return (hPa * 0.0295300).toFixed(2);
    },
```

```
    disconnect: function(e) {
        if (app.peripheral && app.peripheral.id) {
            ble.disconnect(app.peripheral.id, app.showMainPage, app.onError);
        }
    },
    showMainPage: function() {
        mainPage.hidden = false;
        detailPage.hidden = true;
    },
    showDetailPage: function() {
        mainPage.hidden = true;
        detailPage.hidden = false;
    },
    onError: function(reason) {
        navigator.notification.alert(reason, app.showMainPage, "Error");
    }
};

app.initialize();
```

Run the App

Save *index.html* and *index.js* and you're ready to run the application. Go back to your terminal or command prompt and start the server by typing phonegap serve.

```
xvi:weather don(master)$ phonegap serve
[phonegap] starting app server...
[phonegap] listening on 10.0.1.16:3000
[phonegap]
[phonegap] ctrl-c to stop the server
[phonegap]
```

On your phone or tablet, start the PhoneGap Developer App. Verify that the server address listed in the app matches the address from the phonegap serve command. Press Connect. You should see a list of devices offering the Weather Service. Click on your device. You should see the weather details reported on the detail screen (Figure 6-9). Click the disconnect button to go back to the device list.

Figure 6-9   *Left: PhoneGap Developer App. Center: Weather main page. Right: Weather detail page*

The source code for this chapter is available on GitHub (*https://github.com/MakeBluetooth/ ble-weather*).

What's Next?

You combined an Arduino, a BME280 breakout board, and a Bluetooth service to create a weather station that measures temperature, humidity, and pressure.

More sophisticated weather stations measure wind speed, wind direction, rainfall, and dew point. If you'd like, you can expand your weather station with additional sensors. Additional sensors will require new characteristics in the Bluetooth service. This sketch maxes out the available memory on the Arduino Uno, so a more sophisticated weather station may require a different Arduino board with more memory.

The PhoneGap application read and converted the Bluetooth data for display. If you want a more permanent record of the data, you can build a program for your computer that collects the data over Bluetooth and logs it to a file, spreadsheet, or web service.

NeoPixel Lamp | 7

In this chapter, you will build an LED light that can be controlled from an app on your phone.

Hardware

The following hardware is required:

- Arduino Uno
- Bluefruit LE nRF8001 (*http://adafru.it/1697*)
- NeoPixel Ring with 16 LEDs (*http://adafru.it/1463*)
- Rotary encoder (*http://www.adafruit.com/products/377*)
- AC adapter, 9V or 12V

NeoPixels

This project uses NeoPixel LEDs. NeoPixel is Adafruit's name for its individually controllable WS2812 LEDs. Each NeoPixel in a ring can be set to one of 16 million colors. For this project, all the lights will be the same color. NeoPixels can be sensitive to voltage, so it's important to keep them supplied with 5V or less. The NeoPixels are also sensitive to CPU clock timing, but the Adafruit Arduino library handles that for us.

Building the Hardware

Solder wires onto a NeoPixel ring. Twenty-two-gauge solid-core wire works well because you can bend it to adjust the direction of the light. Use red for 5VDC, black for ground, and blue for data input. Leave the data output free. Be careful not to solder the wires to the LED contacts or melt the plastic on the LEDs (see Figure 7-1.)

 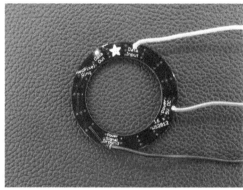

Figure 7-1 *Left: NeoPixel ring. Right: Solder NeoPixel wires*

The NeoPixel lights are very bright. If you have access to a 3D printer, you can print a diffuser using clear filament. Get the *neoring-cover.stl* file from the Camera LED Ring (*http://www.thingiverse.com/thing:235242/#files*) project on Thingiverse (*http://thingiverse.com*). Once you print the ring, snap it onto the NeoPixel ring (Figure 7-2).

 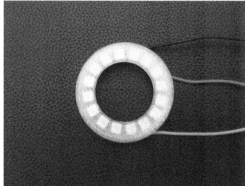

Figure 7-2 *3D-printed NeoPixel cover*

Wire the Bluetooth breakout board into the Arduino Uno board using the standard wiring shown in "Wiring Up the Adafruit Bluefruit LE Module". Connect the NeoPixel ring to the Uno. Put the red wire into 5V, the black wire into GND, and the blue wire into pin 6, as shown in Figure 7-3.

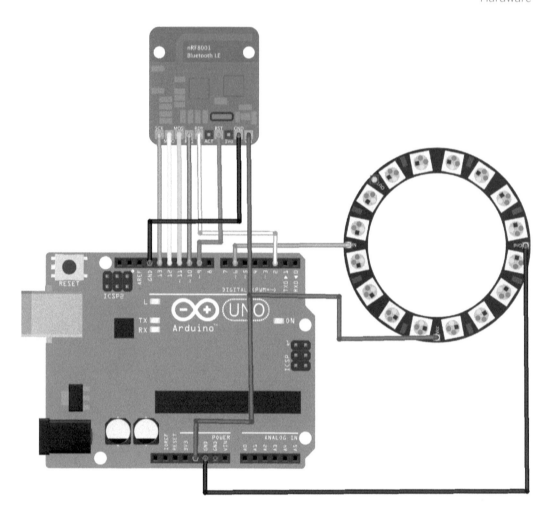

Figure 7-3 Bluetooth NeoPixel diagram

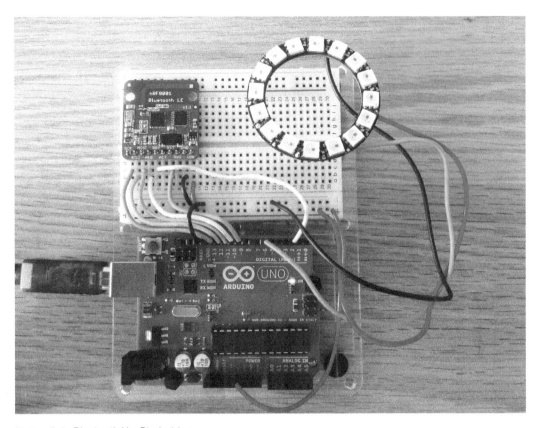

Figure 7-4 *Bluetooth NeoPixel wiring*

Software

Now that you have built the hardware, it is time to write the software. First, you will define a Bluetooth service to control the lights.

LED Service

The LED service (Table 7-1) has three characteristics. The color characteristic allows you to set the color of the LEDs using a 24-bit number, such as 0xFF0000 for red, 0x00FF00 for green, or 0x87CEEB for sky blue. The 24-bit color is built by combining the red, green, and blue values into one number. This is very similar to defining CSS colors (*http://mzl.la/1Sb9pOX*) for web pages.

Table 7-1 *LED Service ccc0*

Characteristic	UUID	Properties	Comment
Color (24-bit)	ccc1	read, write	0 to 0xFFFFFF
Brightness	ccc2	read, write, notify	0 to 0xFF
Switch	ccc3	read, write, notify	1 on, 0 off

The brightness characteristic allows the brightness to be set from off (0) to full brightness (255). Note that the integer 255 is represented as 0xFF in hexadecimal. This is the largest value that can fit in one 8-bit byte.

The power characteristic turns the lights on and off but keeps the Bluetooth radio running. Send 1 to switch the lights on and 0 to turn the lights off.

Programming the Arduino

Our sketch requires two Arduino libraries:

- BLEPeripheral (*https://github.com/sandeepmistry/arduino-BLEPeripheral*)
- Adafruit NeoPixel (*https://github.com/adafruit/Adafruit_NeoPixel*)

Use the Arduino IDE Library Manager to install the libraries by going to Sketch → Include Library → Manage Libraries…. The BLEPeripheral library is probably installed if you've done another project in this book. If not, refer to "Installing the BLE Peripheral Library" for more details. Add the Adafruit NeoPixel library to the Arduino IDE.

Create a new Arduino sketch to control the NeoPixels. Use File → New in the Arduino IDE to create the file. Then use File → Save As to name the sketch NeoPixel. Click the save button.

At the top of the sketch, import the SPI, BLEPeripheral, and NeoPixel libraries.

```
#include <SPI.h>
#include <BLEPeripheral.h>
#include <Adafruit_NeoPixel.h>
```

The BLEPeripheral library can be configured to work with different Bluetooth LE hardware. Define the constants for the Adafruit Bluefruit LE nRF8001 and create an instance of the BLEPeripheral. If you're using different hardware, consult the library documentation (*https://github.com/sandeepmistry/arduino-BLEPeripheral#pinouts*) for the proper settings.

```
#define BLE_REQ 10
#define BLE_RDY 2
#define BLE_RST 9

BLEPeripheral blePeripheral = BLEPeripheral(BLE_REQ, BLE_RDY, BLE_RST);
```

Next, create the Bluetooth Low Energy, based on the information in Table 7-1. The table can be translated into code.

```
BLEService neoPixelService = BLEService("ccc0");

BLECharacteristic colorCharacteristic = BLECharacteristic("ccc1", BLERead | BLEWrite, 3);
BLEDescriptor colorDescriptor = BLEDescriptor("2901", "Color (24-bit)");
BLEUnsignedCharCharacteristic brightnessCharacteristic =
            BLEUnsignedCharCharacteristic("ccc2", BLERead | BLEWrite | BLENotify);
BLEDescriptor brightnessDescriptor = BLEDescriptor("2901", "Brightness");
BLEUnsignedCharCharacteristic switchCharacteristic =
            BLEUnsignedCharCharacteristic("ccc3", BLERead | BLEWrite | BLENotify);
BLEDescriptor switchDescriptor = BLEDescriptor("2901", "Power Switch");
```

The BLEDescriptors in the code aren't required, but they are very helpful. Tools like Light-Blue will display the descriptor in the user interface. It's much easier to understand "Color (24-bit)" than "ccc1."

The NeoPixels also need to be configured. The NeoPixel ring has 16 LEDs, so set NUM BER_PIXELS to 16. If you have a larger or smaller NeoPixel ring or strip, adjust this number. The data wire for the NeoPixels is plugged into pin 6. Set the initial brightness to 25%. After these constants are defined, create an Adafruit_NeoPixel instance.

```
#define NUMBER_PIXELS 16
#define NEO_PIXEL_PIN 6
#define DEFAULT_BRIGHTNESS 0x3F // 25%

Adafruit_NeoPixel pixels = Adafruit_NeoPixel(NUMBER_PIXELS, NEO_PIXEL_PIN, NEO_GRB +
NEO_KHZ800);
```

Create a setup() function. Setup runs once when the Arduino starts the program.

```
void setup() {
}
```

Add a line to start Serial, so we can print debug information and view it using the Serial Monitor in the Arduino IDE. Print a message, Bluetooth Low Energy NeoPixel, so it's easy to see which sketch is running. The F() is a macro that ensures the string is defined in PROGMEM and there's more SRAM for your program to run.

```
Serial.begin(9600);
Serial.println(F("Bluetooth Low Energy NeoPixel"));
```

Set the NEO_PIXEL_PIN to an output pin, and call begin() to start the NeoPixel library.

```
pinMode(NEO_PIXEL_PIN, OUTPUT);
pixels.begin();
```

The blePeripheral needs some configuration. Set the localName and deviceName to assist with device discovery. Set the NeoPixel service to be the advertised BLE service.

```
blePeripheral.setDeviceName("NeoPixels");
blePeripheral.setLocalName("NeoPixels");
blePeripheral.setAdvertisedServiceUuid(neoPixelService.uuid());
```

See "Device Name Versus Local Name" for more information about device name and local name.

In the preamble of the sketch, you created the objects for the Bluetooth Low Energy Service. These objects need to be added into the peripheral to create the service. The service, characteristics, and descriptors are all added as attributes. The order in which they are added is important.

```
blePeripheral.addAttribute(neoPixelService);
blePeripheral.addAttribute(colorCharacteristic);
blePeripheral.addAttribute(colorDescriptor);
blePeripheral.addAttribute(brightnessCharacteristic);
blePeripheral.addAttribute(brightnessDescriptor);
blePeripheral.addAttribute(switchCharacteristic);
blePeripheral.addAttribute(switchDescriptor);
```

When a Bluetooth client changes the value of the color characteristic, the color of the Neo-Pixels should change. Register an event handler on the colorCharacteristic. BLEWritten is a constant. The colorCharacteristicWritten callback function will be defined later.

```
colorCharacteristic.setEventHandler(BLEWritten, colorCharacteristicWritten);
```

The brightness and switch characteristics also need event handlers.

```
brightnessCharacteristic.setEventHandler(BLEWritten, brightnessCharacteristicWritten);
switchCharacteristic.setEventHandler(BLEWritten, switchCharacteristicWritten);
```

Now that the blePeripheral is set up, call begin.

```
blePeripheral.begin();
```

Last, set the LEDs to the default brightness, set the initial color to blue, and call update Lights to light the LEDs.

```
brightnessCharacteristic.setValue(DEFAULT_BRIGHTNESS);
pixels.setBrightness(DEFAULT_BRIGHTNESS);
const unsigned char initialColor[3] = {0x00, 0x00, 0xFF}; // red, green, blue
colorCharacteristic.setValue(initialColor, sizeof(initialColor));

updateLights();
```

This completes the setup function.

The loop() function is called continuously while the Arduino program is running. For this sketch, the loop function simply calls blePeripheral.poll(), which tells the Bluetooth library to do whatever it should be working on.

```
void loop() {
  blePeripheral.poll();
}
```

Next, define the callback functions that handle changes in attribute values. These functions were referenced earlier in setup with characteristic.setEventHandler.

The `colorCharacteristicWritten` function delegates all its work to the `updateLights` function, which will be defined later.

```
void colorCharacteristicWritten(BLECentral& central,
                                BLECharacteristic& characteristic) {
    updateLights();
}
```

The `brightnessCharacteristicWritten` function sets the brightness based on the value of the characteristic, and then calls `updateLights` to ensure the proper color is displayed. Calling `updateLights` ensures the LEDs are on when brightness goes from 0 to a positive value. It also ensures the lights turn on, if they were previously off.

```
void brightnessCharacteristicWritten(BLECentral& central,
                                     BLECharacteristic& characteristic) {
    pixels.setBrightness(brightnessCharacteristic.value());
    updateLights();
}
```

The `switchCharacteristicWritten` function is called when a client writes a new value to the switch characteristic. The `switchCharacteristicWritten` delegates all the work to the `processSwitchChange` function. Creating a separate function allows this logic to be reused later.

```
void switchCharacteristicWritten(BLECentral& central,
                                 BLECharacteristic& characteristic) {
    processSwitchChange();
}
```

The `processSwitchChange` function is slightly more complex. It is possible to turn off the lights by setting the brightness to zero. When the switch is turned on, check if the brightness is zero and reset it to the default brightness. Then call `updateLights` to light the LEDs using the current color and brightness values. When the power switch turns off, set all the pixels to 0, meaning black. This turns off the lights but allows the previous brightness and color values to be retained.

```
void processSwitchChange() {
    if (switchCharacteristic.value() == 1) {
        if (pixels.getBrightness() == 0) {
            brightnessCharacteristic.setValue(DEFAULT_BRIGHTNESS);
            pixels.setBrightness(DEFAULT_BRIGHTNESS);
        }
        // updateLights uses the last color and brightness
        updateLights();
    } else if (switchCharacteristic.value() == 0) {
        // turn all pixels off
        for (int i = 0; i < NUMBER_PIXELS; i++) {
            pixels.setPixelColor(i, 0);
        }
        pixels.show();
    }
}
```

The last function you need to define is updateLights. The updateLights function gets the 24-bit value from the colorCharacteristic and breaks it into the three 8-bit values for the red, green, and blue color components. The Adafruit NeoPixel library creates a 32-bit integer from the red, green, and blue color components. Then, in a loop, each pixel on the ring is set to the desired color. Calling pixel.show() causes the color to display.

```
void updateLights() {
  // get the color array from the characteristic
  const unsigned char* rgb = colorCharacteristic.value();
  uint8_t red = rgb[0];
  uint8_t green = rgb[1];
  uint8_t blue = rgb[2];

  // change the color of the lights
  uint32_t color = pixels.Color(red, green, blue);
  for (int i = 0; i < NUMBER_PIXELS; i++) {
    pixels.setPixelColor(i, color);
  }
  pixels.show();

  // ensure that the switch characteristic is correct
  if (switchCharacteristic.value() == 0 && pixels.getBrightness() > 0) {
    switchCharacteristic.setValue(1); // light is on
  } else if (pixels.getBrightness() == 0 && switchCharacteristic.value() == 1) {
    switchCharacteristic.setValue(0); // light is off
  }
}
```

See Example 7-1 for the complete listing of *BLE_NeoPixel*.

Example 7-1 BLE_NeoPixel.ino

```
#include <SPI.h>
#include <BLEPeripheral.h>
#include <Adafruit_NeoPixel.h>

#define BLE_REQ 10
#define BLE_RDY 2
#define BLE_RST 9

BLEPeripheral blePeripheral = BLEPeripheral(BLE_REQ, BLE_RDY, BLE_RST);

BLEService neoPixelService = BLEService("ccc0");

BLECharacteristic colorCharacteristic =
          BLECharacteristic("ccc1", BLERead | BLEWrite, 3);
BLEDescriptor colorDescriptor = BLEDescriptor("2901", "Color (24-bit)");
BLEUnsignedCharCharacteristic brightnessCharacteristic =
          BLEUnsignedCharCharacteristic("ccc2", BLERead | BLEWrite | BLENotify);
BLEDescriptor brightnessDescriptor = BLEDescriptor("2901", "Brightness");
BLEUnsignedCharCharacteristic switchCharacteristic =
          BLEUnsignedCharCharacteristic("ccc3", BLERead | BLEWrite | BLENotify);
```

```
BLEDescriptor switchDescriptor = BLEDescriptor("2901", "Power Switch");

#define NUMBER_PIXELS 16
#define NEO_PIXEL_PIN 6
#define DEFAULT_BRIGHTNESS 0x3F // 25%

Adafruit_NeoPixel pixels = Adafruit_NeoPixel(NUMBER_PIXELS, NEO_PIXEL_PIN, NEO_GRB +
NEO_KHZ800);

void setup() {
  Serial.begin(9600);
  Serial.println(F("Bluetooth Low Energy NeoPixel"));

  pinMode(NEO_PIXEL_PIN, OUTPUT);
  pixels.begin();

  // set advertised name and service
  blePeripheral.setDeviceName("NeoPixels");
  blePeripheral.setLocalName("NeoPixels");
  blePeripheral.setAdvertisedServiceUuid(neoPixelService.uuid());

  // add service and characteristic
  blePeripheral.addAttribute(neoPixelService);
  blePeripheral.addAttribute(colorCharacteristic);
  blePeripheral.addAttribute(colorDescriptor);
  blePeripheral.addAttribute(brightnessCharacteristic);
  blePeripheral.addAttribute(brightnessDescriptor);
  blePeripheral.addAttribute(switchCharacteristic);
  blePeripheral.addAttribute(switchDescriptor);

  // handlers for when clients change data
  colorCharacteristic.setEventHandler(BLEWritten, colorCharacteristicWritten);
  brightnessCharacteristic.setEventHandler(BLEWritten, brightnessCharacteristicWritten);
  switchCharacteristic.setEventHandler(BLEWritten, switchCharacteristicWritten);

  blePeripheral.begin();

  // initial brightness and color
  brightnessCharacteristic.setValue(DEFAULT_BRIGHTNESS);
  pixels.setBrightness(DEFAULT_BRIGHTNESS);
  const unsigned char initialColor[3] = {0x00, 0x00, 0xFF}; // red, green, blue
  colorCharacteristic.setValue(initialColor, sizeof(initialColor));

  updateLights();
}

void loop() {
  // Tell the bluetooth radio to do whatever it should be working on
  blePeripheral.poll();
}

void colorCharacteristicWritten(BLECentral& central,
                                BLECharacteristic& characteristic) {
  updateLights();
```

```
  }

  void brightnessCharacteristicWritten(BLECentral& central,
                                       BLECharacteristic& characteristic) {
    pixels.setBrightness(brightnessCharacteristic.value());
    updateLights();
  }

  void switchCharacteristicWritten(BLECentral& central,
                                   BLECharacteristic& characteristic) {
    processSwitchChange();
  }

  void processSwitchChange() {
    if (switchCharacteristic.value() == 1) {
      if (pixels.getBrightness() == 0) {
        brightnessCharacteristic.setValue(DEFAULT_BRIGHTNESS);
        pixels.setBrightness(DEFAULT_BRIGHTNESS);
      }
      // updateLights uses the last color and brightness
      updateLights();
    } else if (switchCharacteristic.value() == 0) {
      // turn all pixels off
      for (int i = 0; i < NUMBER_PIXELS; i++) {
        pixels.setPixelColor(i, 0);
      }
      pixels.show();
    }
  }

  void updateLights() {
    // get the color array from the characteristic
    const unsigned char* rgb = colorCharacteristic.value();
    uint8_t red = rgb[0];
    uint8_t green = rgb[1];
    uint8_t blue = rgb[2];

    // change the color of the lights
    uint32_t color = pixels.Color(red, green, blue);
    for (int i = 0; i < NUMBER_PIXELS; i++) {
      pixels.setPixelColor(i, color);
    }
    pixels.show();

    // ensure the switch characteristic is correct
    if (switchCharacteristic.value() == 0 && pixels.getBrightness() > 0) {
      switchCharacteristic.setValue(1); // light is on
    } else if (pixels.getBrightness() == 0 && switchCharacteristic.value() == 1) {
      switchCharacteristic.setValue(0); // light is off
    }
  }
```

Generic Bluetooth Client

Now that we've created the NeoPixel LED peripheral, let's control it using a generic Bluetooth client on your phone. Android owners can use nRF Master Control Panel (*http://bit.ly/1Sb9ySu*). iPhone owners should use LightBlue (*http://bit.ly/1hq3m9j*).

Scan for the lights. Look for a device named NeoPixels.

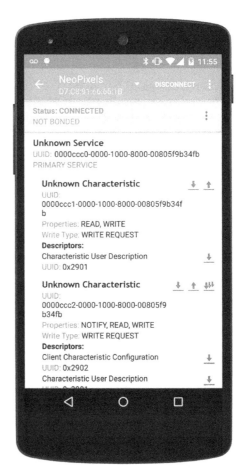

Figure 7-5 *Left: LightBlue. Right: nRF Master Control Panel*

Once you've connected to the NeoPixel device, try changing some data using the LED Service we created. Choose the ccc1 color characteristic. Enter **00ff00** and send. The lights should turn green. Try other colors like red (**ff0000**), magenta (**ff00ff**), or sky blue (**87ceeb**). Experiment and create your own colors.

Connect to the ccc2 brightness characteristic. Set different values. Brightness is a 1-byte value: **ff** is the brightest, **7f** is 50%, and **00** is off.

Connect to the `ccc3` switch characteristic. Toggle the light on and off by setting this characteristic to **01** or **00**.

Building a Phone App

The generic Bluetooth client works well for testing the light and LED Service, but it is a bit cumbersome to dial in the setting you want. Next, you will build an iPhone or Android application to control the light using PhoneGap. For more information on PhoneGap and getting your system set up, refer to "Installing PhoneGap".

This application works similar to the PhoneGap applications in Chapters 4 and 6. When the application starts, it scans for peripherals that are advertising the LED Service (Table 7-1) and presents the user with a list of devices. The user taps on the device they want to control. The software connects and presents the user with a screen to control the color and brightness of the lights.

Create the Project

Open a terminal or cmd prompt on your computer. Change to the directory where you'd like to create the project.

```
cd ~/bluetooth
```

Use the `phonegap` command-line tool to create a new project.

```
phonegap create neopixel com.makebluetooth.neopixel NeoPixel --template blank
```

Open the new project with your text editor.

HTML

The user interface of our application is built with HTML (Example 7-2). The main page displays a list of discovered Bluetooth devices (Figure 7-6). The detail page is used to control the LEDs. Three sliders are used to set the red, green, and blue components of the color. Another slider adjusts the brightness of the light. A checkbox is used to toggle the light on and off (Figure 7-6).

Example 7-2 index.html

```html
<!DOCTYPE html>
<html>
    <head>
        <meta http-equiv="Content-Type" content="text/html; charset=UTF-8" />
        <meta name = "format-detection" content = "telephone=no"/>
        <meta name="viewport" content="initial-scale=1, maximum-scale=1,
         user-scalable=no, width=device-width">
        <link rel="stylesheet" type="text/css" href="css/index.css">
        <title>BLE NeoPixel</title>
    </head>
    <body>
        <div id="connectionScreen">
```

```html
        <h1>BLE NeoPixel</h1>
        <ul id="deviceList">
        </ul>
        <button id="refreshButton">Refresh</button>
    </div>

    <div id="colorScreen">
        <p>Use sliders to set color.</p>
        Red
        <input type="range" min="0" max="255" value="0" id="red"/>
        Green
        <input type="range" min="0" max="255" value="0" id="green"/>
        Blue
        <input type="range" min="0" max="255" value="0" id="blue"/>
        Brightness
        <input type="range" min="0" max="255" value="0" id="brightness" />
        <div>
          <label>
            <input type="checkbox" class="switch" id="powerSwitch" />Power
          </label>
        </div>

        <div id="rgbText">
             
        </div>
        <div id="previewColor">
        </div>

        <button id="disconnectButton">Disconnect</button>
    </div>
    <div id="messageDiv"></div>
    <script type="text/javascript" src="cordova.js"></script>
    <script type="text/javascript" src="js/index.js"></script>
</body>
</html>
```

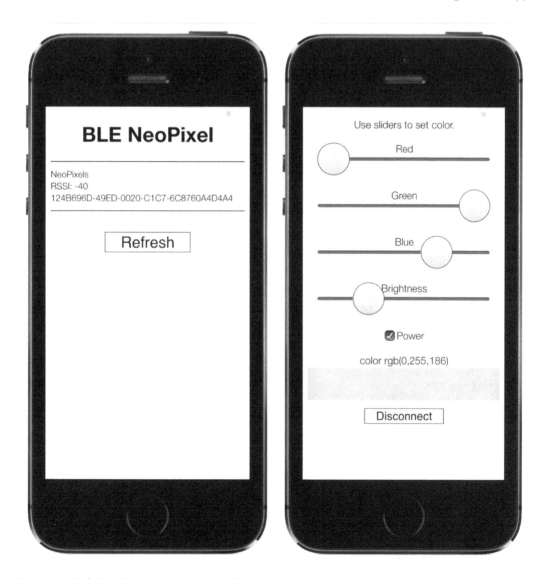

Figure 7-6 *Left: NeoPixel connection screen. Right: NeoPixel color screen*

CSS

The application will work with plain HTML, but it looks better if some CSS is added. Create a directory named *css* under *www*. Copy the CSS from GitHub (*http://bit.ly/1Sb9K4o*) into *www/css/index.css*.

JavaScript

The code for application initialization, Bluetooth scanning, listing devices, and connecting follows the same pattern as the applications from the other projects, so those details will not be explained again. If necessary, refer to the PhoneGap sections of Chapters 4 and 6 for more details.

After the application successfully connects to a Bluetooth device, onConnect is called. The onConnect function switches the view to the color screen, calls syncUI, and registers notifications for the brightness and switch characteristics. The notifications ensure that the phone is notified when the characteristic value changes. The success callback will be called whenever the characteristic value changes. The success callback updates the user interface with the new value. The second part of this chapter will make use of these notifications.

```
// in app.onConnect
ble.startNotification(peripheral.id, LED_SERVICE, BRIGHTNESS, function(buffer) {
    var data = new Uint8Array(buffer);
    brightness.value = data[0];
});
```

The syncUI function ensures the user interface shows the same values as the Arduino. This is done by issuing Bluetooth read requests for the color, brightness, and switch characteristics. The success callbacks processes the data and adjusts the UI. The color is sent as a 24-bit value. A Uint8Array is created from the ArrayBuffer. This allows the data to be accessed one byte at a time. The color data is set to the values for the red, green, and blue range controls. When the value of the range control is set, the user interface is updated to reflect the new value.

```
// in app.syncUI
ble.read(id, LED_SERVICE, COLOR, function(buffer) {
    var data = new Uint8Array(buffer);
    red.value = data[0];
    green.value = data[1];
    blue.value = data[2];
    app.updatePreview();
});
```

The user interface has slider controls for the red, green, and blue components of the desired color. Each of these color channels can be adjusted individually from 0 to 255. The three sliders all have the same event handlers whenever a value changes.

```
red.onchange = app.onColorChange;
green.onchange = app.onColorChange;
blue.onchange = app.onColorChange;
```

The onColorChange event handler updates to the preview color on the screen, and then calls sendColorToArduino.

```
onColorChange: function (evt) {
    app.updatePreview();
    app.sendColorToArduino();
},
```

The color characteristic of the LED Service is 3 bytes. The sendColorToArduino function creates a new Uint8Array of size 3 and adds the red, green, and blue color components. This data is sent to the Arduino using the ble.write API call. The ble.write expects the data to be sent as an ArrayBuffer. The typed `Uint8Array` has a buffer property that gives you the ArrayBuffer. The success and failure callbacks are defined as inline functions. When the write is successful, the status message reports that the color was set. If there was an error, the error will be displayed as a status message.

```
sendColorToArduino: function() {
    var value = new Uint8Array(3);
    value[0] = red.value;
    value[1] = green.value;
    value[2] = blue.value;
    ble.write(app.connectedPeripheral.id, LED_SERVICE, COLOR, value.buffer,
        function() {
            app.setStatus("Set color to " + app.getColor());
        },
        function(error) {
            app.setStatus("Error setting characteristic " + error);
        }
    );
},
```

The brightness and switch controls have similar event handlers. The `onBrightnessChange` and `onSwitchChange` functions each read a value from the HTML control and send the value to the Arduino using `ble.write`.

Example 7-3 has the complete listing of *index.js*.

Example 7-3 *index.js*

```
// LED Service UUIDs
var LED_SERVICE = 'ccc0';
var COLOR = 'ccc1';
var BRIGHTNESS = 'ccc2';
var POWER_SWITCH = 'ccc3';

var app = {
    initialize: function() {
        this.bind();
    },
    bind: function() {
        document.addEventListener('deviceready', this.deviceready, false);
        colorScreen.hidden = true;
    },
    deviceready: function() {

        // wire buttons to functions
        deviceList.ontouchstart = app.connect;
        refreshButton.ontouchstart = app.scan;
        disconnectButton.ontouchstart = app.disconnect;

        red.onchange = app.onColorChange;
        green.onchange = app.onColorChange;
        blue.onchange = app.onColorChange;
```

```
        brightness.onchange = app.onBrightnessChange;

        powerSwitch.onchange = app.onSwitchChange;
        app.scan();
    },
    scan: function(e) {
        deviceList.innerHTML = ""; // clear the list
        app.setStatus("Scanning for Bluetooth Devices...");

        ble.startScan([LED_SERVICE],
            app.onDeviceDiscovered,
            function() { app.setStatus("Listing Bluetooth Devices Failed"); }
        );

        // stop scan after 5 seconds
        setTimeout(ble.stopScan, 5000, app.onScanComplete);

    },
    onDeviceDiscovered: function(device) {
        var listItem, rssi;

        console.log(JSON.stringify(device));
        listItem = document.createElement('li');
        listItem.dataset.deviceId = device.id;
        if (device.rssi) {
            rssi = "RSSI: " + device.rssi + "<br/>";
        } else {
            rssi = "";
        }
        listItem.innerHTML = device.name + "<br/>" + rssi + device.id;
        deviceList.appendChild(listItem);

        var deviceListLength = deviceList.getElementsByTagName('li').length;
        app.setStatus("Found " + deviceListLength +
                    " device" + (deviceListLength === 1 ? "." : "s."));
    },
    onScanComplete: function() {
        var deviceListLength = deviceList.getElementsByTagName('li').length;
        if (deviceListLength === 0) {
            app.setStatus("No Bluetooth Peripherals Discovered.");
        }
    },
    connect: function (e) {
        app.setStatus("Connecting...");
        var deviceId = e.target.dataset.deviceId;
        console.log("Requesting connection to " + deviceId);
        ble.connect(deviceId, app.onConnect, app.onDisconnect);
    },
    disconnect: function(event) {
        app.setStatus("Disconnecting...");
        ble.disconnect(app.connectedPeripheral.id, app.onDisconnect);
    },
    onConnect: function(peripheral) {
        app.connectedPeripheral = peripheral;
```

```javascript
            connectionScreen.hidden = true;
            colorScreen.hidden = false;
            app.setStatus("Connected.");
            app.syncUI();

            // notifications update the UI if the values change on the light
            ble.startNotification(peripheral.id, LED_SERVICE, BRIGHTNESS, function(buffer) {
                var data = new Uint8Array(buffer);
                brightness.value = data[0];
            });

            ble.startNotification(peripheral.id, LED_SERVICE, POWER_SWITCH,
            function(buffer) {
              var data = new Uint8Array(buffer);
              powerSwitch.checked = data[0] !== 0;
            });

    },
    onDisconnect: function() {
        connectionScreen.hidden = false;
        colorScreen.hidden = true;
        app.setStatus("Disconnected.");
    },
    syncUI: function() {
        // read values from light and update the phone UI
        var id = app.connectedPeripheral.id;
        ble.read(id, LED_SERVICE, COLOR, function(buffer) {
            var data = new Uint8Array(buffer);
            red.value = data[0];
            green.value = data[1];
            blue.value = data[2];
            app.updatePreview();
        });
        ble.read(id, LED_SERVICE, BRIGHTNESS, function(buffer) {
            var data = new Uint8Array(buffer);
            brightness.value = data[0];
        });
        ble.read(id, LED_SERVICE, POWER_SWITCH, function(buffer) {
          var data = new Uint8Array(buffer);
          powerSwitch.checked = data[0] !== 0;
        });
    },
    onColorChange: function (evt) {
        app.updatePreview();
        app.sendColorToArduino();
    },
    updatePreview: function() {
        var c = app.getColor();
        rgbText.innerText = "color rgb(" + c  + ")";
        previewColor.style.backgroundColor = "rgb(" + c + ")";
    },
    getColor: function () {
        // returns a string of red, green, blue values
        var color = [];
```

```
            color.push(red.value);
            color.push(green.value);
            color.push(blue.value);
            return color.join(',');
    },
    sendColorToArduino: function() {
        var value = new Uint8Array(3);
        value[0] = red.value;
        value[1] = green.value;
        value[2] = blue.value;
        ble.write(app.connectedPeripheral.id, LED_SERVICE, COLOR, value.buffer,
            function() {
                app.setStatus("Set color to " + app.getColor());
            },
            function(error) {
                app.setStatus("Error setting characteristic " + error);
            }
        );
    },
    onBrightnessChange: function(evt) {
      // user adjusted the brightness with the slider
      var value = new Uint8Array(1);
      value[0] = brightness.value;
      ble.write(app.connectedPeripheral.id, LED_SERVICE, BRIGHTNESS, value.buffer,
            function() {
                app.setStatus("Set brightness to " +  brightness.value);
            },
            function(error) {
                app.setStatus("Error setting characteristic " + error);
            }
        );
    },
    onSwitchChange: function(evt) {
      // the user toggled the power switch
      var value = new Uint8Array(1);
      if (powerSwitch.checked) {
        value[0] = 1; // turn on
      } else {
        value[0] = 0; // turn off
      }
      ble.write(app.connectedPeripheral.id, LED_SERVICE, POWER_SWITCH, value.buffer,
            function() {
                app.setStatus("Set switch to " +  value[0]);
            },
            function(error) {
                app.setStatus("Error setting characteristic " + error);
            }
        );
    },
    timeoutId: 0,
    setStatus: function(status) {
        if (app.timeoutId) {
            clearTimeout(app.timeoutId);
        }
```

```
        messageDiv.innerText = status;
        app.timeoutId = setTimeout(function() { messageDiv.innerText = ""; }, 4000);
    }
};

app.initialize();
```

 On Android 4.3 and 4.4, scan filtering is broken. Often you cannot filter scan results by UUID. If you are not seeing any devices while scanning, try passing an empty array [] to ble.scan instead of filtering for [LED_SER VICE].

Run the App

Save *index.html* and *index.js* and you're ready to run the application. Go back to your terminal or command prompt and start the server by typing phonegap serve.

```
xvi:neopixel don(master)$ phonegap serve
[phonegap] starting app server...
[phonegap] listening on 10.0.1.16:3000
[phonegap]
[phonegap] ctrl-c to stop the server
[phonegap]
```

On your phone or tablet, start the PhoneGap Developer App. Verify that the server address listed in the app matches the address from the phonegap serve command. Press Connect. You should see a list of devices offering the LED Service. Click on your device. You should see the page that allows you to control the NeoPixel lights on the Arduino (Figures 7-6 and 7-7). Click the disconnect button to go back to the device list.

Figure 7-7 *Left: PhoneGap Developer App. Center: NeoPixel main screen. Right: NeoPixel color screen*

Enhancements

Now that you can control the NeoPixel lights with your phone, let's look at enhancing the project by adding some physical controls ("Physical Switch and Dimmer") and embedding the Arduino and NeoPixels into a lamp ("Lamp").

Physical Switch and Dimmer

Having to take your phone out of your pocket to turn a light on and off can be a pain. Let's add an external power switch and dimmer knob. This physical switch will work similar to the switch characteristic. The light will turn on and off but the Bluetooth radio will stay on. The dimmer knob will adjust the brightness and can also be used to turn the light off by turning the brightness all the way down.

A rotary encoder is a component that emits a pulse as it rotates. The rotary encoder (Figure 7-8) looks like a potentiometer, but the implementation is different. A potentiometer varies the resistance as it rotates through a range. A potentiometer will have a specific resistance for each position. The rotary encoder continuously rotates; there are no end stops. When the encoder spins, we use the Arduino library to count the number of clicks and translate this into a brightness. If the brightness goes below zero, we set it to zero. If the brightness goes over the maximum of 0xFF, it gets reset to the maximum. The rotary encoder works well here since the brightness can be set via Bluetooth or the encoder. The rotary encoder from Adafruit also includes a push-button switch.

Figure 7-8 *Rotary encoder*

One side of the rotary encoder has three pins. The center pin is a ground, and the side pins are used to measure the clicks as the encoder is rotated. The other side of the rotary encoder has two pins. These pins are connected to a push-button switch. When the encoder is depressed, the switch is closed.

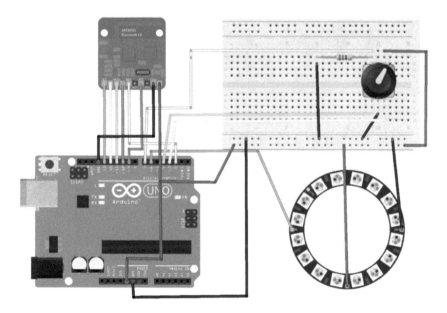

Figure 7-9 *NeoPixel lamp diagram*

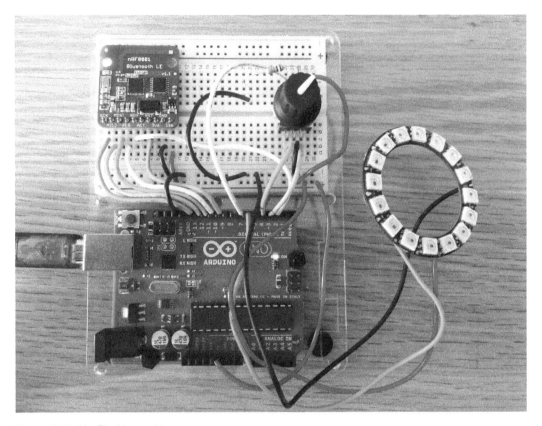

Figure 7-10 *NeoPixel lamp wiring*

Once the rotary encoder and switch are wired in (Figures 7-8 and 7-9), the sketch needs to be adjusted. The sketch needs to handle the physical input from the new hardware and keep this in sync with the Bluetooth characteristics.

Create a new sketch by making a copy. Use File → Save As to save the *BLE_NeoPixel* sketch as *BLE_NeoPixel_Lamp*.

Install the Encoder library (*http://www.pjrc.com/teensy/td_libs_Encoder.html*) using the Arduino Library Manager. Include the Encoder library at the beginning of the sketch.

```
#include <Encoder.h>
```

After the existing DEFAULT_BRIGHTNESS constant, define MAX_BRIGHTNESS and set the value to 0xFF.

```
#define NUMBER_PIXELS 16
#define NEO_PIXEL_PIN 6
#define DEFAULT_BRIGHTNESS 0x3F // 25%
#define MAX_BRIGHTNESS 0xFF
```

The encoder needs pins 3 and 4 to read clicks. `BRIGHTNESS_PER_CLICK` is a multiplier so the dimmer works faster; otherwise, it requires too many revolutions to go from off to the brightest setting.

```
#define PIN_ENCODER_A 3
#define PIN_ENCODER_B 4
#define BRIGHTNESS_PER_CLICK 3
```

The power button will attach to pin 7.

```
#define BUTTON_PIN 7
```

Add some variables to track the encoder value and button state.

```
uint8_t encoderValue;
int buttonState = 0;
```

Add a variable, `previousMillis`, to track the last time the encoder state was read. Add `in terval` to track the interval between reads. These variables are used to limit how often the rotary encoder and button state are read.

```
long previousMillis = 0;  // stores the last time sensor was read
long interval = 100;      // interval at which to read sensor (milliseconds)
```

Create an instance of the encoder just before the setup function.

```
Encoder encoder(PIN_ENCODER_A, PIN_ENCODER_B);
```

The setup function needs two minor updates. Set the button pin to an input.

```
pinMode(BUTTON_PIN, INPUT);
pinMode(NEO_PIXEL_PIN, OUTPUT);
```

Set the encoder brightness to the default brightness. This initializes the encoder value so it is in sync with the LEDs and the brightness characteristic. The `setEncoderBrightness` function will be defined later.

```
brightnessCharacteristic.setValue(DEFAULT_BRIGHTNESS);
pixels.setBrightness(DEFAULT_BRIGHTNESS);
setEncoderBrightness(DEFAULT_BRIGHTNESS);
const unsigned char initialColor[3] = {0x00, 0x00, 0xFF}; // red, green, blue
colorCharacteristic.setValue(initialColor, sizeof(initialColor));
```

The `loop` function needs to read the encoder value and button value. It is better to add a delay between reads, rather than reading the values continuously. Many Arduino sketches will use a `delay(100)` to slow down the processing of the `loop` function. This works effectively in some non-Bluetooth sketches, but it's not recommended with Bluetooth because the radio can't work during the delay (*http://www.arduino.cc/en/Reference/Delay*).

An alternative is to use millis (*http://www.arduino.cc/en/Reference/Millis*). The `millis` function will tell us the number of milliseconds since the Arduino board started running the current program. Track the millis when the encoder state is read and ensure there is a delay of at least 100 milliseconds before it is read again.

```
void loop() {
  blePeripheral.poll();

  if (millis() - previousMillis > interval) {
  readEncoder();
  readButton();
  previousMillis = millis();
 }

}
```

The `brightnessCharacteristicWritten` function needs to set the encoder value in addition to setting the pixel brightness.

```
void brightnessCharacteristicWritten(BLECentral& central, BLECharacteristic&
characteristic) {
  pixels.setBrightness(brightnessCharacteristic.value());
  setEncoderBrightness(brightnessCharacteristic.value());
  updateLights();
}
```

The `processSwitchChange` function also needs to set the encoder brightness.

```
void processSwitchChange() {
  if (switchCharacteristic.value() == 1) {
    if (pixels.getBrightness() == 0) {
        brightnessCharacteristic.setValue(DEFAULT_BRIGHTNESS);
        setEncoderBrightness(DEFAULT_BRIGHTNESS);
        pixels.setBrightness(DEFAULT_BRIGHTNESS);
    }
    // ...
  }
  // ...
}
```

When the brightness is changed via Bluetooth, the encoder value needs to be updated. Create a helper function so you set the encoder using a brightness value.

```
void setEncoderBrightness(uint8_t brightness) {
  encoder.write(brightness / BRIGHTNESS_PER_CLICK);
}
```

Now, implement the `readEncoder` function. The raw value is read from the encoder. If the value changed since the last time it was read, ensure that it's not below the minimum or over the maximum. Once you know the value is in range, scale it up to a brightness but multiplying by `BRIGHTNESS_PER_CLICK`. Set the value of the brightness characteristic. Set the brightness of the LEDs. Lastly, call `updateLights` to ensure the LEDs display the proper color.

```
void readEncoder() {
  long val = encoder.read();

  if (val != encoderValue) { // value changed
    // don't go below 0
```

```
      if (val < 0) {
        val = 0;
        encoder.write(val);
      }
      // don't go above max
      if (val > MAX_BRIGHTNESS/BRIGHTNESS_PER_CLICK) {
        val = MAX_BRIGHTNESS/BRIGHTNESS_PER_CLICK;
        encoder.write(val);
      }

      encoderValue = val;
      Serial.println(encoderValue);

      //sync the characteristic
      uint8_t brightness = encoderValue * BRIGHTNESS_PER_CLICK;
      brightnessCharacteristic.setValue(brightness);
      pixels.setBrightness(brightness);
      updateLights();
    }
  }
```

The physical button on the rotary encoder can be used to toggle the light on and off. When the button is pressed, the reading is HIGH. Flip the value of the switchCharacteristic, then call processSwitchChanged to set the actual state of the lights.

```
  void readButton() {
    // read the state of the pushbutton value:
    buttonState = digitalRead(BUTTON_PIN);

    // check if the pushbutton is pressed.
    if (buttonState == HIGH) {

      if (switchCharacteristic.value() > 0) {
        // light was on, turn it off
        switchCharacteristic.setValue(0);
      } else {
        // light was off, turn it on
        switchCharacteristic.setValue(1);
      }

      processSwitchChange();
      delay(200);
    }
  }
```

Compile and run the sketch. Now you can turn the lights on and off with the physical switch, as well as adjust the brightness using the rotary encoder. You'll still need your phone to adjust the color using Bluetooth.

See Example 7-4 for the complete listing of *BLE_NeoPixel_Lamp.ino*.

Example 7-4 *BLE_NeoPixel_Lamp.ino*

```
// BLE NeoPixel
//
// Bluefruit LE http://adafru.it/1697
// NeoPixels LEDs http://adafru.it/1463
// arduino-BLEPeripheral https://github.com/sandeepmistry/arduino-BLEPeripheral.git
// Adafruit NeoPixel Driver https://github.com/adafruit/Adafruit_NeoPixel
// Encoder Library http://www.pjrc.com/teensy/td_libs_Encoder.html

#include <SPI.h>
#include <BLEPeripheral.h>
#include <Adafruit_NeoPixel.h>
#include <Encoder.h>

// See BLE Peripheral documentation for setting for your hardware
// https://github.com/sandeepmistry/arduino-BLEPeripheral#pinouts
#define BLE_REQ 10
#define BLE_RDY 2
#define BLE_RST 9

BLEPeripheral blePeripheral = BLEPeripheral(BLE_REQ, BLE_RDY, BLE_RST);

BLEService neoPixelService = BLEService("ccc0");

BLECharacteristic colorCharacteristic =
          BLECharacteristic("ccc1", BLERead | BLEWrite, 3);
BLEDescriptor colorDescriptor = BLEDescriptor("2901", "Color (24-bit)");
BLEUnsignedCharCharacteristic brightnessCharacteristic =
          BLEUnsignedCharCharacteristic("ccc2", BLERead | BLEWrite | BLENotify);
BLEDescriptor brightnessDescriptor = BLEDescriptor("2901", "Brightness");
BLEUnsignedCharCharacteristic switchCharacteristic =
          BLEUnsignedCharCharacteristic("ccc3", BLERead | BLEWrite | BLENotify);
BLEDescriptor switchDescriptor = BLEDescriptor("2901", "Power Switch");

#define NUMBER_PIXELS 16
#define NEO_PIXEL_PIN 6
#define DEFAULT_BRIGHTNESS 0x3F // 25%
#define MAX_BRIGHTNESS 0xFF

#define PIN_ENCODER_A 3
#define PIN_ENCODER_B 4
#define BRIGHTNESS_PER_CLICK 3

#define BUTTON_PIN 7

uint8_t encoderValue;
int buttonState = 0;

long previousMillis = 0;  // stores the last time sensor was read
long interval = 100;      // interval at which to read sensor (milliseconds)

Adafruit_NeoPixel pixels = Adafruit_NeoPixel(NUMBER_PIXELS, NEO_PIXEL_PIN, NEO_GRB +
```

```
NEO_KHZ800);
Encoder encoder(PIN_ENCODER_A, PIN_ENCODER_B);

void setup() {
  Serial.begin(9600);
  Serial.println(F("Bluetooth Low Energy NeoPixel Lamp"));

  // initialize the pushbutton pin as an input:
  pinMode(BUTTON_PIN, INPUT);

  pinMode(NEO_PIXEL_PIN, OUTPUT);
  pixels.begin();

  // set advertised name and service
  blePeripheral.setDeviceName("NeoPixels");
  blePeripheral.setLocalName("NeoPixels");
  blePeripheral.setAdvertisedServiceUuid(neoPixelService.uuid());

  // add service and characteristic
  blePeripheral.addAttribute(neoPixelService);
  blePeripheral.addAttribute(colorCharacteristic);
  blePeripheral.addAttribute(colorDescriptor);
  blePeripheral.addAttribute(brightnessCharacteristic);
  blePeripheral.addAttribute(brightnessDescriptor);
  blePeripheral.addAttribute(switchCharacteristic);
  blePeripheral.addAttribute(switchDescriptor);

  // handlers for when clients change data
  colorCharacteristic.setEventHandler(BLEWritten, colorCharacteristicWritten);
  brightnessCharacteristic.setEventHandler(BLEWritten, brightnessCharacteristicWritten);
  switchCharacteristic.setEventHandler(BLEWritten, switchCharacteristicWritten);

  blePeripheral.begin();

  // initial brightness and color
  brightnessCharacteristic.setValue(DEFAULT_BRIGHTNESS);
  pixels.setBrightness(DEFAULT_BRIGHTNESS);
  setEncoderBrightness(DEFAULT_BRIGHTNESS);
  const unsigned char initialColor[3] = {0x00, 0x00, 0xFF}; // red, green, blue
  colorCharacteristic.setValue(initialColor, sizeof(initialColor));

  updateLights();
}

void loop() {
  // Tell the bluetooth radio to do whatever it should be working on
  blePeripheral.poll();

  // limit how often we read the rotary encoder and button
  if (millis() - previousMillis > interval) {
    readEncoder();
    readButton();
    previousMillis = millis();
  }
```

```
    }

    void colorCharacteristicWritten(BLECentral& central,
                                    BLECharacteristic& characteristic) {
      updateLights();
    }

    void brightnessCharacteristicWritten(BLECentral& central,
                                         BLECharacteristic& characteristic) {
      pixels.setBrightness(brightnessCharacteristic.value());
      setEncoderBrightness(brightnessCharacteristic.value());
      updateLights();
    }

    void switchCharacteristicWritten(BLECentral& central,
                                     BLECharacteristic& characteristic) {
      processSwitchChange();
    }

    void processSwitchChange() {
      if (switchCharacteristic.value() == 1) {
        if (pixels.getBrightness() == 0) {
            brightnessCharacteristic.setValue(DEFAULT_BRIGHTNESS);
            setEncoderBrightness(DEFAULT_BRIGHTNESS);
            pixels.setBrightness(DEFAULT_BRIGHTNESS);
        }
        // updateLights uses the last color and brightness
        updateLights();
      } else if (switchCharacteristic.value() == 0) {
        // turn all pixels off
        for (int i = 0; i < NUMBER_PIXELS; i++) {
          pixels.setPixelColor(i, 0);
        }
        pixels.show();
      }
    }

    void updateLights() {
      // get the color array from the characteristic
      const unsigned char* rgb = colorCharacteristic.value();
      uint8_t red = rgb[0];
      uint8_t green = rgb[1];
      uint8_t blue = rgb[2];

      // change the color of the lights
      uint32_t color = pixels.Color(red, green, blue);
      for (int i = 0; i < NUMBER_PIXELS; i++) {
        pixels.setPixelColor(i, color);
      }
      pixels.show();

      // ensure the switch characteristic is correct
      if (switchCharacteristic.value() == 0 && pixels.getBrightness() > 0) {
```

```
      switchCharacteristic.setValue(1); // light is on
    } else if (pixels.getBrightness() == 0 && switchCharacteristic.value() == 1) {
      switchCharacteristic.setValue(0); // light is off
    }
}

void setEncoderBrightness(uint8_t brightness) {
  encoder.write(brightness / BRIGHTNESS_PER_CLICK);
}

void readEncoder() {
  long val = encoder.read();

  if (val != encoderValue) { // value changed
    // don't go below 0
    if (val < 0) {
      val = 0;
      encoder.write(val);
    }
    // don't go above max
    if (val > MAX_BRIGHTNESS/BRIGHTNESS_PER_CLICK) {
      val = MAX_BRIGHTNESS/BRIGHTNESS_PER_CLICK;
      encoder.write(val);
    }

    encoderValue = val;
    Serial.println(encoderValue);

    //sync the characteristic
    uint8_t brightness = encoderValue * BRIGHTNESS_PER_CLICK;
    brightnessCharacteristic.setValue(brightness);
    pixels.setBrightness(brightness);
    updateLights();
  }
}

void readButton() {
  // read the state of the pushbutton value:
  buttonState = digitalRead(BUTTON_PIN);

  // check if the pushbutton is pressed.
  if (buttonState == HIGH) {

    if (switchCharacteristic.value() > 0) {
      // light was on, turn it off
      switchCharacteristic.setValue(0);
    } else {
      // light was off, turn it on
      switchCharacteristic.setValue(1);
    }

    processSwitchChange();
    delay(200);
```

```
    }
  }
```

The source code for *BLE_NeoPixel_Lamp.ino* is also available in the book's source code repository (*http://bit.ly/1WKHdbX*).

Lamp

The LED lights are nice and the Arduino works great on our workbench, but it might not fit in with your home decor. Let's embed the electronics into a common household lamp. The Ikea Lampan is a good place to start since it is inexpensive and has a flat base to easily fit the electronics (Figure 7-11). You might want to choose a nicer lamp for your project.

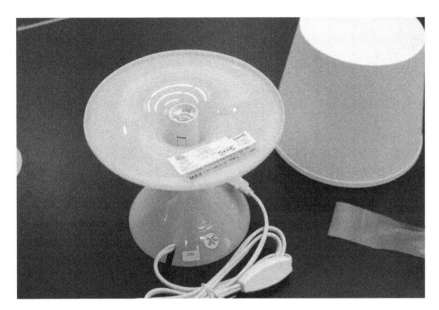

Figure 7-11   *Ikea lamp*

Cut the power cord under the lamp. Pry out the light socket and pull the cord through. There is not enough room to feed the power adapter and wires through the base. Unscrew and remove the plastic piece that holds the two parts of the base together. This will create a larger opening, but it will also cause the base to split into two parts (Figure 7-12). Hot glue the lamp base back together.

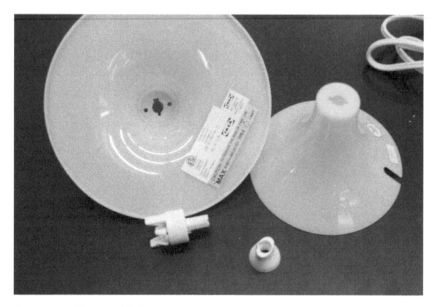

Figure 7-12 *Disassembled lamp*

Solder long wires onto the rotary encoder. You'll need enough wire to thread through the neck of the lamp. Melt a hole in the plastic and mount the rotary encoder in the lamp base (Figure 7-13). Run the wires through the center of the lamp. Run the 9V power adapter through the center of the lamp (Figure 7-14).

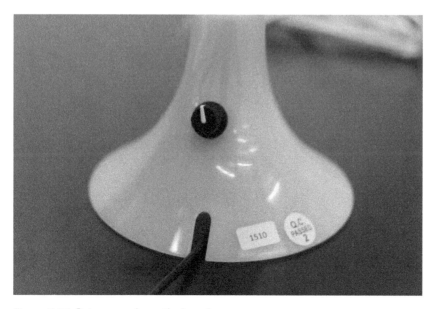

Figure 7-13 *Rotary encoder on the lamp base*

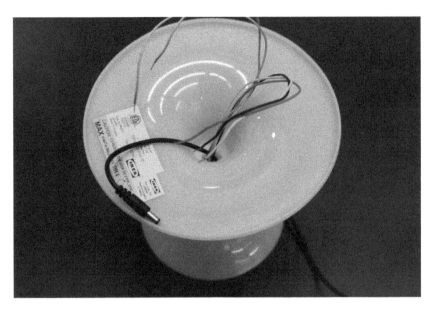

Figure 7-14 *Wires going through the base*

Place the Arduino on top of the base. Attach the rotary encoder to the Uno. Plug the power adapter into the barrel jack. Test the LED with your phone. If everything works, snap the cover on the lamp.

Figure 7-15 *Arduino installed in the lamp*

Now that the lamp is assembled, you can use the phone app to set the color, control the brightness, and switch the lamp on and off (Figure 7-16). When your phone is not available, use the knob on the lamp to switch the power and control the brightness.

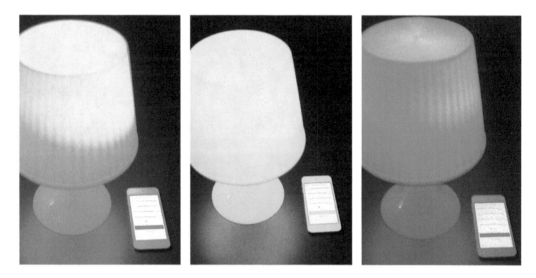

Figure 7-16 *Colors!*

SensorTag Remote

8

The SensorTag Development Kit is a piece of hardware from Texas Instruments built to showcase TI's Bluetooth Low Energy chip. The SDK is intended for developing wireless sensor applications. The SensorTag contains a number of sensors, including temperature, humidity, pressure, accelerometer, gyroscope, and magnetometer. It also includes two buttons. The original version was the CC2541 SensorTag Development Kit. In 2015, TI introduced a next generation of the SDK called the CC2650 SensorTag.

This project uses the buttons on the SensorTag. When a button is pressed, a program on your computer will receive a Bluetooth notification and act as if someone pressed the left or right arrow key on the keyboard. You can use this remote control to navigate through a Powerpoint or Keynote presentation.

This project is based on an example (*http://bit.ly/1SbdrXQ*) Tom Igoe wrote for the ITP classes (*http://itp.nyu.edu/itp/*) at NYU. We've replaced AppleScript in the original version so it can run on OS X, Linux, and maybe even Windows.

Hardware

The CC2650 SensorTag Development Kit can be purchased for $29 plus shipping directly from Texas Instruments (*http://www.ti.com/tool/cc2650stk*), or from distributors such as DigiKey. TI stopped selling the CC2541 SensorTag (*http://www.ti.com/tool/cc2541dk-sensor*), but you may still be able to find it from other suppliers. I think the button layout is awkward on the newer CC2650 and prefer the older CC2451 for this project (Figure 8-1).

Figure 8-1 *CC2541 and CC2650 SensorTag Development Kits*

If you don't have a SensorTag, you can build an Arduino version to use for this project. See "Arduino Simple Key Service" at the end of the chapter for more info.

The Bluetooth Low Energy services available on the SensorTag are documented on the Texas Instruments wiki (*http://bit.ly/1kQHyTq*). For this project, we will use the Simple Key Service (*http://bit.ly/1SbdxhS*) (Table 8-1).

Table 8-1 *Simple Key Service FFE0*

Characteristic	UUID	Properties	Comment
Button State	FFE1	notify	Bit 1 - right key, Bit 0 - left key

When using the remote, your computer is the Bluetooth Low Energy central device and the SensorTag is the peripheral. The computer initiates the connection to the SensorTag. Once connected, the computer subscribes to be notified whenever a button is pressed. After the initial connection, the SensorTag (or server) initiates all the communication to the computer (or client). When the state of a button change, the SensorTag notifies the computer.

You don't need to write any code to run on the SensorTag. It comes with the Simple Key Service wired to the buttons. You will write a program that runs on your computer using Node.js and the noble (*http://bit.ly/1NHoaVR*) library. See "Installing Node.js" for more information on setting your computer up to run Node.js. You might also want to check "Making Sure Your Machine Has Bluetooth LE" to ensure your computer is ready.

Create the Project

Open a terminal or cmd prompt on your computer. Change to the directory where you'd like to create the project.

```
cd ~/bluetooth
```

Create a new directory, **sensor-tag**, for the project.

```
mkdir sensor-tag
cd sensor-tag/
```

This project needs the noble (*https://www.npmjs.com/package/noble*) library. Use npm to install it. Ensure that you run the command in the sensor-tag project directory.

```
npm install noble
```

See "Installing Node.js" if you need more information on Node.js, or "Installing Libraries with npm" for npm libraries. If you see errors while installing noble check the noble documentation (*https://github.com/sandeepmistry/noble/wiki/Getting-started*).

SensorTag and Noble

Now that your project is set up and noble is installed, create a new file named *sensortag.js* in the *sensor-tag* directory. Open the file in a text editor.

Require the noble library at the beginning of *sensortag.js*.

```
var noble = require('noble');
```

When the application starts, noble fires the stateChange event with the current state of the Bluetooth adapter. If the Bluetooth adapter is powered on, the application starts scanning for Bluetooth Low Energy devices. No arguments are passed to noble.startScanning so all Bluetooth Low Energy devices will be discovered. Register a state change listener with noble.

```
noble.on('stateChange', function(state) {
  if (state === 'poweredOn') {
    noble.startScanning(); // any service UUID
  } else {
    console.log('Please power-on the Bluetooth Adapter.');
  }
});
```

Discovery of the SensorTag is a bit tricky. Ideally, we want to look for a device that is advertising the Simple Key Service, `FFE0`. Unfortunately the CC2541 SensorTag does not include any services in the advertising data. (The newer CC2650 advertises service `AA10`.) Instead of filtering while scanning, the code looks at the peripheral's local name to decide if it's a SensorTag or not.

Add another callback for noble's `discover` event. The function will be called each time a Bluetooth Low Energy peripheral is discovered. Use a regular expression to see if the name contains the word `Sensor`. If it does, assume a SensorTag was found. Stop the scan and call `connectAndSetUpSensorTag`.

```
noble.on('discover', function(peripheral) {

  var localName = peripheral.advertisement.localName;

  // find SensorTag based on local name
  if (localName && localName.match(/Sensor/)) {
    noble.stopScanning();
    console.log('Attempting to connect to ' + localName);
    connectAndSetUpSensorTag(peripheral);
  }
});
```

Create a new function named `connectAndSetUpSensorTag`. The connect logic could have been nested inside the callback for `noble.on`, but it is often clearer to create named functions, avoiding deeply nested logic.

The `connectAndSetUpSensorTag` function calls `peripheral.connect`. Like most of the other API calls, the connection happens asynchronously. The callback is invoked once the connection is complete. Check the error parameter to ensure there weren't any problems connecting.

The peripheral object has methods for discovering services and discovering characteristics. Calling these methods means another series of nested callbacks. `discoverSomeServicesAndCharacteristics` is a convenience function that combines service and characteristic discovery into one callback. Rather than look for all services and characteristics, the UUIDs for the Simple Key Service and Button Characteristic are passed in to limit discovery to only the attributes we need.

The `onDisconnect` function is attached to the peripheral's disconnect event.

```
function connectAndSetUpSensorTag(peripheral) {

  peripheral.connect(function(error) {
    console.log('Connected to ' + peripheral.advertisement.localName);
    if (error) {
      console.log('There was an error connecting ' + error);
      return;
    }

    var serviceUUIDs = ['FFE0'];
```

```
  var characteristicUUIDs = ['FFE1'];

  peripheral.discoverSomeServicesAndCharacteristics(
    serviceUUIDs, characteristicUUIDs, onServicesAndCharacteristicsDiscovered);
  });

  // attach disconnect handler
  peripheral.on('disconnect', onDisconnect);
}
```

Create the event handler for the disconnect event. For now, print a log message to the console.

```
function onDisconnect() {
  console.log('Peripheral disconnected!');
}
```

The onServicesAndCharacteristicsDiscovered function is called after noble discovers the service and characteristic data for the connected peripheral. Like most Node.js callbacks, the first thing to do is check for errors. Then, get the button characteristic, FFE1, from the characteristic array. The discovery was filtered to find only one characteristic, so we take the first element of the array.

Subscribe to be notified when the characteristic value changes by calling characteristic.notify(true).

Set an inline event handler for the characteristic notification state changes. If the characteristic is notifying, tell the user the remote is ready. This step isn't strictly necessary but makes the app a little nicer.

Lastly, add an event handler for the characteristic's data event. This is the important event that notifies us of changes to the characteristic's data. The data event is called any time the characteristic's data is read or there is a notification from the peripheral. This application only uses notification. Delegate this event handler to the onCharacteristicData function.

```
function onServicesAndCharacteristicsDiscovered(error, services, characteristics) {

  if (error) {
    console.log('Error discovering services and characteristics ' + error);
    return;
  }

  var characteristic = characteristics[0];

  // subscribe for notifications
  characteristic.notify(true);

  // called when notification state changes
  characteristic.on('notify', function(isNotifying) {
    if (isNotifying) {
      console.log('SensorTag remote is ready');
    }
  });
```

```
  // called when the data changes
  characteristic.on('data', onCharacteristicData);
}
```

We've finally reached the point where we get data from the button. Noble will call onChar acteristicData whenever the SensorTag's button is pressed. Data is passed to the onChar acteristicData function as a Node.js Buffer (*https://nodejs.org/api/buffer.html*).

Button information is encoded in two bits of data. The lowest bit represents the right button state and the second bit represents the left button state. 0 means the button is not pressed. 1 means the button is pressed. Table 8-2 shows how we can interpret the results that are returned by the SensorTag.

Table 8-2 *Data values for Simple Key Service*

Hex	Int	Meaning
0x00	0	No buttons are pressed
0x01	1	Right button is pressed
0x10	2	Left button is pressed
0x11	3	Both buttons is pressed

Write a method to translate the button data and display a console message.

```
function onCharacteristicData(data, isNotification) {
  switch (data[0]) {
  case 0:
    console.log("No buttons are pressed");
    break;
  case 1:
    console.log("Right button is pressed");
    break;
  case 2:
    console.log("Left button is pressed");
    break;
  case 3:
    console.log("Both buttons are pressed");
    break;
  default:
    console.log("Error " + data[0]);
  }
}
```

The complete listing for *sensortag.js* is in Example 8-1. You can also refer to GitHub (*https://github.com/makebluetooth/sensor-tag-remote*).

Example 8-1 *sensortag.js*

```
var noble = require('noble');

noble.on('stateChange', function(state) {
  if (state === 'poweredOn') {
    noble.startScanning(); // any service UUID
  } else {
    console.log('Please power-on the Bluetooth Adapter.');
  }
});

noble.on('discover', function(peripheral) {

  var localName = peripheral.advertisement.localName;

  // find SensorTag based on local name
  if (localName && localName.match(/Sensor/)) {
    noble.stopScanning();
    console.log('Attempting to connect to ' + localName);
    connectAndSetUpSensorTag(peripheral);
  }
});

function connectAndSetUpSensorTag(peripheral) {

  peripheral.connect(function(error) {
    console.log('Connected to ' + peripheral.advertisement.localName);
    if (error) {
      console.log('There was an error connecting ' + error);
      return;
    }

    var serviceUUIDs = ['FFE0'];
    var characteristicUUIDs = ['FFE1'];

    peripheral.discoverSomeServicesAndCharacteristics(
    serviceUUIDs, characteristicUUIDs, onServicesAndCharacteristicsDiscovered);
  });

  // attach disconnect handler
  peripheral.on('disconnect', onDisconnect);
}

function onDisconnect() {
  console.log('Peripheral disconnected!');
}

function onServicesAndCharacteristicsDiscovered(error, services, characteristics) {

  if (error) {
    console.log('Error discovering services and characteristics ' + error);
    return;
```

```
    }

    var characteristic = characteristics[0];

    // subscribe for notifications
    characteristic.notify(true);

    // called when notification state changes
    characteristic.on('notify', function(isNotifying) {
      if (isNotifying) {
        console.log('SensorTag remote is ready');
      }
    });

    // called when the data changes
    characteristic.on('data', onCharacteristicData);
}

function onCharacteristicData(data, isNotification) {
  switch (data[0]) {
  case 0:
    console.log("No buttons are pressed");
    break;
  case 1:
    console.log("Right button is pressed");
    break;
  case 2:
    console.log("Left button is pressed");
    break;
  case 3:
    console.log("Both buttons are pressed");
    break;
  default:
    console.log("Error " + data[0]);
  }
}
```

Launch your program from a command prompt by typing node *sensortag.js*. Press the power button on your SensorTag to wake it up. Wait for the "SensorTag remote is ready" message. Press the left and right buttons on the SensorTag and watch the output. Hit Ctrl-C to exit the program. See Example 8-2 for sample output.

Example 8-2 *Sample output from sensortag.js*

```
$ node sensortag.js
Attempting to connect to TI BLE Sensor Tag
Connected to SensorTag
SensorTag remote is ready
Left button is pressed
No buttons are pressed
Right button is pressed
```

```
No buttons are pressed
Right button is pressed
Both buttons are pressed
No buttons are pressed
^C
```

 You'll need to run these examples with sudo on Linux. Noble needs root privileges to access the Bluetooth adapter.

SensorTag Remote

Now that you know SensorTag works and can interpret the button presses, you're ready to create the remote control.

The robotjs (*https://www.npmjs.com/package/robotjs*) library will be used to send key codes from our application to the operating system. The program on your computer mimics a user typing on a keyboard, which means whatever program that is in focus will receive the keystrokes.

Install robotjs using npm. Ensure that you run the command in the sensor-tag project directory.

```
npm install robotjs
```

Only a few changes are needed to make the program into a remote control. Copy *sensor-tag.js* to a new file named *remote.js*. Require robotjs at the top of the file.

```
var noble = require('noble');
var robot = require('robotjs');
```

Replace the existing onCharacteristicData function with this new shorter version. This version only handles two cases, the left and right buttons. Keystrokes are sent using robot.keyTap.

```
function onCharacteristicData(data, isNotification) {
  switch (data[0]) {
  case 1:
    console.log('right');
    robot.keyTap('right');
    break;
  case 2:
    console.log('left');
    robot.keyTap('left');
    break;
  }
}
```

Use Node.js to launch *remote.js* from a command prompt. Wait for the SensorTag to connect. (Hit the power button if necessary.)

```
$ node remote.js
Attempting to connect to TI BLE Sensor Tag
Connected to SensorTag
SensorTag remote is ready
```

Open a Keynote or PowerPoint file, enter presentation mode, and use the SensorTag buttons to advance the slides. Note that this also works with web-based presentation software like *reveal.js* (*http://lab.hakim.se/reveal-js/*) or even Google Slides.

See Example 8-3 for the complete program listing. The code is also available from GitHub (*https://github.com/makebluetooth/sensor-tag-remote*).

Example 8-3 *remote.js*

```
var noble = require('noble');
var robot = require('robotjs');

noble.on('stateChange', function(state) {
  if (state === 'poweredOn') {
    noble.startScanning(); // any service UUID
  } else {
    console.log('Please power-on the Bluetooth Adapter.');
  }
});

noble.on('discover', function(peripheral) {

  var localName = peripheral.advertisement.localName;

  // find SensorTag based on local name
  if (localName && localName.match(/Sensor/)) {
    noble.stopScanning();
    console.log('Attempting to connect to ' + localName);
    connectAndSetUpSensorTag(peripheral);
  }
});

function connectAndSetUpSensorTag(peripheral) {

  peripheral.connect(function(error) {
    console.log('Connected to ' + peripheral.advertisement.localName);
    if (error) {
      console.log('There was an error connecting ' + error);
      return;
    }

    var serviceUUIDs = ['FFE0'];
    var characteristicUUIDs = ['FFE1'];

    peripheral.discoverSomeServicesAndCharacteristics(
    serviceUUIDs, characteristicUUIDs, onServicesAndCharacteristicsDiscovered);
  });
```

```
  // attach disconnect handler
  peripheral.on('disconnect', onDisconnect);
}

function onDisconnect() {
  console.log('Peripheral disconnected!');
}

function onServicesAndCharacteristicsDiscovered(error, services, characteristics) {

  if (error) {
    console.log('Error discovering services and characteristics ' + error);
    return;
  }

  var characteristic = characteristics[0];

  // subscribe for notifications
  characteristic.notify(true);

  // called when notification state changes
  characteristic.on('notify', function(isNotifying) {
    if (isNotifying) {
      console.log('SensorTag remote is ready');
    }
  });

  // called when the data changes
  characteristic.on('data', onCharacteristicData);
}

function onCharacteristicData(data, isNotification) {
  switch (data[0]) {
  case 1:
    console.log('right');
    robot.keyTap('right');
    break;
  case 2:
    console.log('left');
    robot.keyTap('left');
    break;
  }
}
```

A Simpler Version

Using noble, you wrote custom code to connect to the SensorTag's Simple Key Service and handle notifications on the button characteristic. A lot of the code was handling the low-level details of device discovery, connecting, service discovery, and characteristic discov-

ery. The important logic for the remote control ended up in the small onCharacteristicDa
ta function.

If you want to use the keys to do something else, you can copy the file and write a new
onCharacteristicData function. Sometimes it makes sense to write a device-specific library that uses noble. A device-specific library can hide the discovery and connection logic
behind a simpler API. The node-sensortag (*https://github.com/sandeepmistry/node-sensor
tag*) library provides higher-level abstraction that can make our code even simpler.

Install sensortag using npm. Ensure that you run the command in the sensor-tag project
directory.

```
npm install sensortag
```

Create a new file, *simple.js*, and type in the code from Example 8-4. This code does the
same thing as *remote.js* but with much less code.

Example 8-4 *simple.js*

```
var SensorTag = require('sensortag');
var robot = require('robotjs');

SensorTag.discover(function(sensorTag) {
  console.log("Found " + sensorTag);

  sensorTag.connectAndSetUp(function(error) {
    sensorTag.notifySimpleKey();
  });

  sensorTag.on('simpleKeyChange', function(left, right) {
    if (right) {
      robot.keyTap('right');
    } else if (left) {
      robot.keyTap('left');
    }
  });

  sensorTag.on('disconnect', function() {
    console.log('SensorTag disconnected!');
  });

});
```

Run this version using node *simple.js*. It should behave exactly like *remote.js*.

Next Steps

The SensorTag remote control sends arrow keys when buttons are pressed. But don't stop
there—you can modify this code to do other things when the buttons are pressed. You

could use robotjs to send different keys or write text, but you're not restricted to robotjs. You can run any code to respond to the button notifications. Use the buttons to run shell scripts, open web pages, launch applications, make web service calls, or launch rockets. The sensortag node library also allows easy access and a nice API to the other sensors on the SensorTag.

Arduino Simple Key Service

If you don't have a TI SensorTag for this chapter, you could buy one, or you can build some hardware with an Arduino that provides the same Simple Key Service. Start with the basic Arduino and Adafruit nRF8001 setup used in the other chapters. Add two push buttons and a few wires as shown in Figure 8-2.

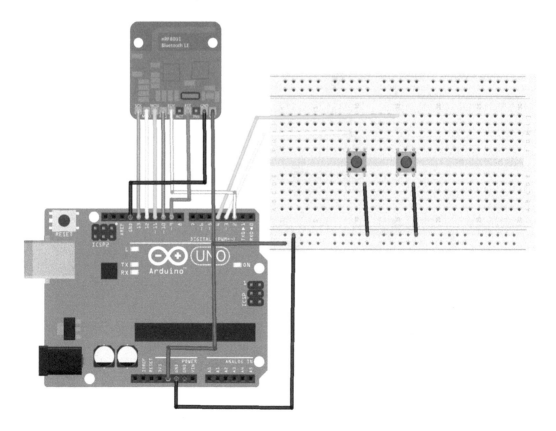

Figure 8-2 *Wiring for fake sensor tag*

Use the Arduino IDE to load the code from Example 8-5 onto your hardware.

Example 8-5 *FakeSensorTag.ino*

```
// TI SensorTag - Simple Key Service
// http://processors.wiki.ti.com/index.php/SensorTag_User_Guide#Simple_Key_Service

#include <SPI.h>
#include <BLEPeripheral.h>

// Adafruit nRF8001
#define BLE_REQ 10
#define BLE_RDY 2
#define BLE_RST 9

#define LEFT_BUTTON  3
#define RIGHT_BUTTON 4

uint8_t buttonState;
uint8_t leftButtonState;
uint8_t rightButtonState;

long lastTime = 0;
uint8_t interval = 20;

BLEPeripheral blePeripheral = BLEPeripheral(BLE_REQ, BLE_RDY, BLE_RST);
BLEService simpleKeyService = BLEService("FFE0");
BLEUnsignedCharCharacteristic characteristic =
            BLEUnsignedCharCharacteristic("FFE1", BLENotify);

void setup() {
  Serial.begin(9600);
  Serial.println(F("SensorTag"));

  pinMode(LEFT_BUTTON, INPUT_PULLUP);
  pinMode(RIGHT_BUTTON, INPUT_PULLUP);

  blePeripheral.setLocalName("SensorT4g");
  blePeripheral.setDeviceName("SensorT4g");

  blePeripheral.addAttribute(simpleKeyService);
  blePeripheral.addAttribute(characteristic);

  blePeripheral.begin();
}

void loop() {
  blePeripheral.poll();

  if (millis() - lastTime > interval) {
    checkButtons();
  }
}

void checkButtons() {
```

```
    leftButtonState = digitalRead(LEFT_BUTTON);
    rightButtonState = digitalRead(RIGHT_BUTTON);

    // using a pullup resistor so LOW means button is pressed
    // assume LOW = 0 and HIGH = 1

    // flip the bits
    rightButtonState = rightButtonState ^ 1;
    leftButtonState = leftButtonState ^ 1;

    // left button is bit 2, shift
    leftButtonState = leftButtonState << 1;

    buttonState = leftButtonState + rightButtonState;

    if (buttonState != characteristic.value()) {
      Serial.print("Setting characteristic to ");
      Serial.println(buttonState, HEX);
      characteristic.setValue(buttonState);
    }

}
```

HID over GATT

9

We use a *Human Interface Device* (HID) every time we interact with a desktop or laptop computer. Keyboards and mice are the most common types of HID device. HID's original definition (*http://bit.ly/1MXzNr0*), uses the USB standard to send and receive data. The next generation of HID devices uses the Bluetooth (*http://bit.ly/1MXzOey*) standard to send and receive data.

The Generic Attribute Profile (GATT) describes how Bluetooth LE transfers data among devices. The *HID over GATT Profile* (HOGP) defines how to create input and output HID devices using BLE.

More information on the HID over GATT Profile can be found on the Bluetooth Developer Portal in the Technology Overview pages (*http://bit.ly/1SbcWwU*).

HOGP and BLEPeripheral

The BLEPeripheral library provides an additional BLEHIDPeripheral API to ease the creation of HOGP devices.

BLEPeripheral has the following HID APIs built in:

BLEKeyboard
> Used for sending standard keyboard key presses

BLEMouse
> Used for sending mouse movement and click inputs

BLEMultimedia
> Used for sending multimedia key presses, such as volume up/down, mute/unmute

BLESystemControl
> Used for sending system-control key presses, such as power and sleep

Users can also implement other custom HID types, such as a joystick, in their sketches if needed.

These HID types can be added to a BLEHIDPeripheral much in the same way that attributes (services, characteristics, descriptors) are added to a BLEPeripheral.

BLEHIDPeripheral extends the BLEPeripheral API by adding the HID Service shown in Table 9-1, along with its characteristics and descriptors. They are required by the HOGP specification.

Using HOGP devices also requires pairing/bonding. Pairing is required to encrypt communications between the peripheral and centra, so eavesdroppers can't spy on your inputs (key strokes) using a BLE sniffer. The BLEHIDPeripheral API manages bonding data by storing it in the Arduino's EEPROM so that it is persisted even when power is lost (unlike RAM).

During pairing, the BLE central and peripheral exchange encryption keys. BLEPeripheral uses *Just Works pairing*. Instead of using a device-specific pairing code, Just Works uses a Temporary Key (TK) that is 16 bytes in size, with all bytes set to zero. The TK is used to calculate a Short-Term Key (STK) to encrypt communications, which is followed by a process to exchange a Long-Term Key (LTK). Both the central and peripheral store the LTK so that communications can continue to be encrypted in the future.

More information on the BLE pairing process can be found at "Bluetooth Low Energy SMP Pairing" (*http://bit.ly/1MXzPPE*).

Table 9-1 *HID Service (UUID 0x1812)*

Attribute	UUID	Properties	Notes
HID Service	0x1812		
HID Information Characteristic	0x2a4a	read	
HID Control Point Characteristic	0x2a4c	write without response	
HID Report Characteristics	0x2a4d	read, notify	One per HID device type
HID Report Reference Descriptor	0x2908		One per HID device type

Volume Knob

To see how this works, let's build a volume knob using a rotary encoder and Arduino. The volume knob will be compatible with iOS 8, Android 5, and OS X 10.10 devices. The volume knob will allow users to increase and decrease the volume on their paired device, as well as to mute and unmute it.

Hardware

The following hardware is required:

- Arduino Uno (*http://www.arduino.cc/en/Main/ArduinoBoardUno*)
- Adafruit Bluefruit LE nRF8001 (*http://www.adafruit.com/products/1697*)
- Breadboard
- Jumper wires
- Rotary encoder with built-in button (Adafruit (*http://www.adafruit.com/products/377*) or Sparkfun (*https://www.sparkfun.com/products/9117*))
- Knob for rotary encoder (optional)

Wiring

The wiring for the Bluefruit LE module is the same as in previous chapters; see "Wiring Up the Adafruit Bluefruit LE Module" for more details.

To wire the rotary encoder, place it in the middle of the breadboard with the side that has two pins facing the top of the breadboard. Now connect the button pins of the rotary encoder by connecting a wire from the top-left pin to pin 5 on the Arduino, and connecting the top-right pin to ground. Then connect the rotary encoder side by connecting a wire from the bottom-left pin to pin 4 on the Arduino, the middle-bottom pin to ground, and the bottom-right pin to pin 3 on the Arduino.

When complete, the wiring will look as it does in Figure 9-1. See Figure 9-2 for a photo of the wire sketch.

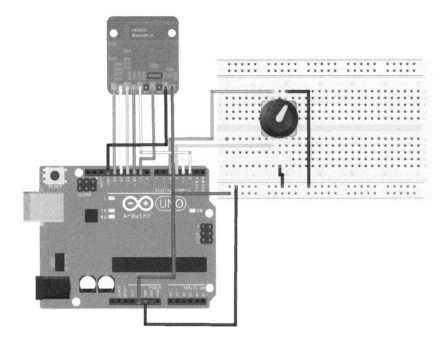

Figure 9-1 *Wiring diagram of HID over GATT volume knob sketch*

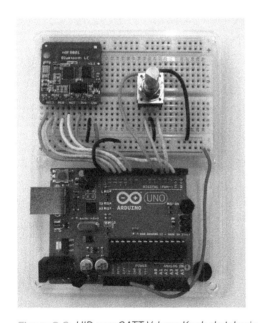

Figure 9-2 *HID over GATT Volume Knob sketch wired*

Arduino Library Setup

We'll be using the following Arduino libraries:

- BLEPeripheral (*https://github.com/sandeepmistry/arduino-BLEPeripheral*)

- PJRC.com Teensy Encoder Library

See "Installing the BLE Peripheral Library" for instructions on setting up the BLEPeripheral library if you haven't already done so.

Setting up the Encoder library

Follow these steps to set up the Encoder library:

1. Open the Arduino IDE.

2. Go to Sketch → Include Library → Manage Libraries…

3. Search for Encoder.

4. Select Encoder by Paul Stoffregen.

5. Click the Install button.

6. Close the Library Manager using the Close button.

Testing the Rotary Encoder

The Encoder library includes a basic example sketch, as shown in Example 9-1. Let's try it out:

1. Go to File → Examples → Encoder → Basic.

2. Update the pin numbers for the encoder to 3 and 4 to match the wiring we set up earlier.

3. Load the sketch onto your Arduino.

4. Open the Serial Monitor and ensure the baud rate is set to 9600.

5. Turn the knob on the rotary encoder.

Example 9-1 Basic.ino sketch

```
/* Encoder Library - Basic Example
 * http://www.pjrc.com/teensy/td_libs_Encoder.html
 *
 * This example code is in the public domain.
 */

#include <Encoder.h> ❶

// Change these two numbers to the pins connected to your encoder.
```

```
//    Best Performance: both pins have interrupt capability
//    Good Performance: only the first pin has interrupt capability
//    Low Performance:  neither pin has interrupt capability
Encoder myEnc(3, 4); ❷
//    avoid using pins with LEDs attached

void setup() {
  Serial.begin(9600);
  Serial.println("Basic Encoder Test:");
}

long oldPosition  = -999;

void loop() {
  long newPosition = myEnc.read(); ❸
  if (newPosition != oldPosition) { ❹
    oldPosition = newPosition;
    Serial.println(newPosition);
  }
}
```

❶   Include the Encoder library.

❷   Create an *Encoder* instance, making sure to update the pin numbers to 3 and 4.

❸   Read the encoder position.

❹   Compare it to the previously read position. If the position has changed, store and print the new value.

The value will:

- Increase when the knob is turned clockwise.

- Decrease when the knob is turned counterclockwise.

Implementing the Volume Knob

Now that we are familiar with the rotary encoder API, let's combine it with the BLEHIDPer ipheral API to create a volume knob.

Let's create a new sketch in the Arduino IDE: File → New.

```
void setup() {
  // put your setup code here, to run once:

}

void loop() {
  // put your main code here, to run repeatedly:

}
```

Now we need to import the libraries we will be using at the start of the sketch.

```
#include <SPI.h>
#include <BLEHIDPeripheral.h>
#include <BLEMultimedia.h>

#include <Encoder.h>
```

SPI is needed by BLEHIDPeripheral, so it is imported first. We use BLEHIDPeripheral instead of BLEPeripheral because it provides an easier interface to build HIDD over GATT peripherals; it uses BLEPeripheral internally. The last import is the Encoder library we used earlier in the chapter.

Next, let's define the constants to match the wiring we completed earlier.

```
#define BLE_REQ    10
#define BLE_RDY    2
#define BLE_RST    9

#define BUTTON_PIN 5

#define ENC_RIGHT_PIN 3
#define ENC_LEFT_PIN  4
```

The first three are used for the Adafruit Bluefruit LE board, followed by the encoder button pin and then the other encoder pins.

Now let's create the object instances and variables we need for the rest of the sketch. We'll need a BLEHIDPeripheral instance to create a BLE peripheral, a BLEMultimedia instance to send multimedia inputs to the central, an Encoder instance, and a variable to keep track of the button state.

```
BLEHIDPeripheral bleHIDPeripheral = BLEHIDPeripheral(BLE_REQ, BLE_RDY, BLE_RST);
BLEMultimedia bleMultimedia;

Encoder encoder(ENC_RIGHT_PIN, ENC_LEFT_PIN);

int buttonState;
```

Next, we move on to the setup function. We will configure the Serial object for output at 9600 baud. Set the BUTTON_PIN as an input in pullup mode. In pullup mode, the Arduino's internal pullup resistor is used, so there is no need to wire an external one on the breadboard.

```
void setup() {
  Serial.begin(9600);

  pinMode(BUTTON_PIN, INPUT_PULLUP);
}
```

We then need to set an initial value of 0 for the encoder instance by using the write method.

```
void setup() {
  // ...

  encoder.write(0);
}
```

Now we will configure the bleHIDPeripheral instance. We'll set the local name that the peripheral advertises to HID Volume. Then we will add our bleMultimedia instances as a HID device. This will automatically create and add new HID Report Characteristics (UUID 0x2a4d) and an HID Report Reference Descriptor (UUID 0x2908) to the peripheral.

```
void setup() {
  // ...

  bleHIDPeripheral.setLocalName("HID Volume");
  bleHIDPeripheral.addHID(bleMultimedia);

  bleHIDPeripheral.begin();
}
```

The final items remaining in setup are to start the bleHIDPeripheral by calling its begin method, and to print a message to the Serial port to indicate that the HID peripheral is set up.

```
void setup() {
  // ...

  bleHIDPeripheral.begin();

  Serial.println(F("BLE HID Volume Knob"));
}
```

If you will be using the volume knob with an Android device, you will also need to include the following line before bleHIDPeripheral.begin() in setup:

```
  bleHIDPeripheral.setReportIdOffset(1);
```

The default report id offset is 0, though most Android devices do not work when it is used so the report id offset needs to be changed to 1.

Now let's implement the loop function of the sketch. We first need to check if the peripheral has a central connected. If is does, we will print the central's address to the serial port.

```
void loop() {
  BLECentral central = bleHIDPeripheral.central();

  if (central) {
    Serial.print(F("Connected to central: "));
    Serial.println(central.address());
  }
}
```

While the central is connected, we need to poll the button and encoder for input events. We call a function named pollInputs, which we will implement shortly.

```
void loop() {
  // ...

  if (central) {
    // ...

    while (bleHIDPeripheral.connected()) {
      pollInputs();
    }
  }
}
```

When the central disconnects, let's also print a message to the serial port.

```
void loop() {
  // ...

  if (central) {
    // ...

    Serial.print(F("Disconnected from central: "));
    Serial.println(central.address());
  }
}
```

Now let's implement the pollInputs function we called from loop earlier. We'll have it call two functions: pollButton for the button and pollEncoder for the encoder.

```
void pollInputs() {
  pollButton();

  pollEncoder();
}
```

The pollButton function will read the current value of the button pin using the digital Read function and compare it to the previously stored value in buttonState. If the value is different, the button has either been pressed or released, and the new value is stored. When pressed, the value will be LOW, so we can send the mute key value MMKEY_MUTE to the central using the write method on the bleMultimedia instance.

```
void pollButton() {
  int tempButtonState = digitalRead(BUTTON_PIN);

  if (tempButtonState != buttonState) {
    buttonState = tempButtonState;

    if (buttonState == LOW) {
      Serial.println(F("Mute"));
      bleMultimedia.write(MMKEY_MUTE);
    }
  }
}
```

In the `pollEncoder`, we first read the current value from the `encoder`. If it is non-zero, the encoder has been rotated. When the value is above `0`, the encoder has been rotated clockwise, which corresponds to a volume-up action. We will map this to the volume-up key value `MMKEY_VOL_UP`, and again use the `write` method on `bleMultimedia` to send the value. When the encoder state is less than zero, it has been turned counterclockwise. This action will be mapped to the volume-down key `MMKEY_VOL_DOWN`. Then we need to reset the encoder's current value to `0`.

```
void pollEncoder() {
  int encoderState = encoder.read();

  if (encoderState != 0) {
    if (encoderState > 0) {
      Serial.println(F("Volume up"));
      bleMultimedia.write(MMKEY_VOL_UP);
    } else {
      Serial.println(F("Volume down"));
      bleMultimedia.write(MMKEY_VOL_DOWN);
    }
    encoder.write(0);
  }
}
```

We can't use the absolute position of the encoder, as we have no way of knowing what the current volume level of the central is. Also, the HID APIs only allows us to send key-press events and not absolute volume levels.

Now you can load the sketch on your Arduino, and then follow the steps in Appendix A to pair the HID peripheral with your iOS, Android, or OS X device.

Once paired successfully, you will be able to control the volume of the device using the knob. You will also be able to toggle muting by pressing the button built into the rotary encoder.

You'll notice that it will turn the volume up or down by more than a notch. We can add some debounce logic to correct this by polling the inputs every 100ms. Update the beginning of the sketch with the following items: a constant for the poll interval, `INPUT_POLL_IN TERVAL`, and a variable named `lastInputPollTime` to store the last input poll time.

```
// ...

#define ENC_RIGHT_PIN 3
#define ENC_LEFT_PIN  4

#define INPUT_POLL_INTERVAL 100

// ...

int buttonState;
unsigned long lastInputPollTime = 0;
```

```
void setup() {

// ...
```

We can then modify the `pollInputs` function to check if the last poll time has been over 100ms. If it has, both the button and encoder inputs are polled. Then we update the `lastInputPollTime` variable. The `millis` function is used for time tracking, and returns the number of milliseconds the sketch has been running.

```
void pollInputs() {
  if (millis() - lastInputPollTime > INPUT_POLL_INTERVAL) {
    pollButton();

    pollEncoder();

    lastInputPollTime = millis();
  }
}
```

The HID peripheral can only be paired with a single device at a time, but we don't yet have a way to clear the pairing data. Let's update the sketch to add this feature. We'll clear the pairing data if the button is pressed on start. This will clear the pairing information stored in the EEPROM.

```
void setup() {
  // ...

  pinMode(BUTTON_PIN, INPUT_PULLUP);
  buttonState = digitalRead(BUTTON_PIN);

  if (buttonState == LOW) {
    Serial.println(F("BLE HID Peripheral - clearing bond data"));

    bleHIDPeripheral.clearBondStoreData();
  }

  encoder.write(0);

  // ...
}
```

To clear the bonding data, hold the rotary encoder button down, and then press the reset button on the Arduino Uno. You will see a message displayed on the Arduino IDE serial monitor when pairing the data has been cleared successfully.

Whenever you clear the bond store of the peripheral, be sure to remove the device from the previously paired device. Instructions for this can be found in Appendix A.

If you encounter issues connecting or pairing with the volume knob peripheral, clear the bond store of the peripheral and unpair it from all previously paired peripherals.

See the HID_volume.ino (*http://bit.ly/1Sbddjq*) example sketch included with the BLEPeripheral library for a full code listing.

Conclusion

In this chapter, we created an Arduino-based BLE volume knob that can be used with a BLE-equipped iOS or Android device, as well as a Mac. We used the HID over GATT APIs included with the BLEPeripheral library, and the PJRC.com Teensy Encoder Library.

You can use the other HID over GATT APIs included with the BLEPeripheral library to create other HOGP devices, such as a numeric keypad or joystick.

Beacons 10

BLE beacons broadcast data using the Generic Access Profile (GAP). A BLE observer does not need to connect to beacons to receive the broadcasted data, it only needs to scan for advertisements.

Beacons are intended to be used for proximity-based applications. For example, your smartphone can use the signal strength of a detected beacon to estimate how close you are to it, and react accordingly by alerting you or providing more relevant content to the beacon you are closest to. The closer you are to a beacon, the more accurate the proximity estimate becomes. Keep in mind that the proximity is just an estimate based on the received signal strength of the beacons; interference from other devices and objects can radically change this value depending on the environment.

What You'll Need

We'll use the following hardware to create beacons and detect them:

- One of the following:
  - Bluetooth-4.0-enabled Mac
  - Linux PC (includes Raspberry Pi or Beagle Bone Black) with a Bluetooth 4.0 adapter, such as the Bluetooth 4.0 USB Module (*https://www.adafruit.com/products/1327*) available on Adafruit
- Bluetooth-4.0-enabled smartphone or tablet running iOS or Android (for testing)

Node.js, PhoneGap, and bleno also need to be set up; see "Installing Node.js", "Installing PhoneGap", and "Platform Tools".

iBeacon

In June 2013, at the annual Worldwide Developers Conference (WWDC) in San Francisco, Apple quietly announced iBeacon. It was not mentioned during the keynote, and only appeared on a single slide about "some other features in the SDK" for iOS 7. At the time, for hardware folks, it was one of the most prominent features of the latest Apple iOS. The high-tech community is always looking for new features that allow apps to do new things.

iBeacon is a technology that allows you to add real-world context to smartphone applications, such as proximity-based alerts or content in applications. It has been integrated into iOS since version 7, both inside the Core Location and the Passkit frameworks, to enable indoor micro-location and geofencing.

What Data Does an iBeacon Advertise?

The iBeacon standard uses the optional manufacturer data (Extended Inquiry Response data type 0xFF) section of the GAP advertising specification to broadcast 25 bytes of data. More information on the manufacturer data section and GAP can be found in the Bluetooth Core Specification.

Here is an example of the data advertised followed by a description of each part:

```
4C00 02 15 B9407F30F5F8466EAFF925556B57FE6D ED4E 8931 B6
```

- The first two bytes are the Apple Company Identifier (little-endian) 0x004C.

- The third byte has a value of 2, which specifies that the data type is iBeacon.

- The fourth byte, 0x15, specifies the remaining data length, which is 21 (0x15) bytes.

- The next 16 bytes contain the iBeacon UUID, B9407F30-F5F8-466E-AFF9-25556B57FE6D.

- The two bytes after the iBeacon UUID are the iBeacon Major (big-endian); i.e., 0xED4E, 60750.

- The two bytes after the iBeacon Major are the iBeacon Minor (big-endian); i.e., 0x8931, 35121.

- The final byte is the measured received signal strength indication (RSSI) at 1 meter away; i.e., 0xB6, -74. You need to take its 2's complement to convert the byte value into a signed number.

Three of the properties create the beacon's identity. These are:

UUID

> This is a property that is unique to each company; in most use cases, the same UUID would be given to all beacons deployed by a company (or group).

Major

> The property you use to specify a related set of beacons (e.g., all the beacons in one store would share the same Major value).

Minor

> The property that you use to specify a particular beacon in a location.

The measured RSSI at 1 meter away is used by the central to estimate proximity to the beacon. This value is used with the RSSI measured by the central when it detects the iBeacon's advertisement. A beacon with a lower (more negative) value further away than a beacon with a larger (less negative) RSSI value.

Building and Detecting a Beacon

Let's create an iBeacon with Node.js and use a smartphone to detect it.

 The Nordic nRF8001 chip used by the Adafruit Bluefruit LE module is only capable of advertising 20 bytes of manufacturer data; 25 bytes are needed to create an iBeacon, so we cannot use this module for it. See "Hardware Suggestions" for Nordic nRF51822-based boards that are capable of advertising as an iBeacon.

We'll use the `node-bleacon` module using Node.js. It uses `bleno` for the BLE layer.

The source code and documentation for `node-bleacon` can be found on GitHub (*http://bit.ly/1NHpl7z*).

First, make a new directory for the project, and change the current directory to it:

```
$ mkdir make-bluetooth-ibeacons
$ cd make-bluetooth-ibeacons
```

Next, install the `bleacon` module from npm:

```
$ npm install bleacon
```

This command will create a `node_modules` folder in the current directory and pull the `bleacon` module, along with its dependencies including `bleno`, down from npm.

Now create a new file called *advertise.js* and open it using your favorite text editor.

At the start of the file, `require` the `bleacon` module:

```
var Bleacon = require('bleacon');
```

Next, set up some variables to store the iBeacon's UUID, major, minor, and measure-powered values:

```
var uuid = 'E2C56DB5-DFFB-48D2-B060-D0F5A71096E0';
var major = 1;
```

```
var minor = 2;
var measuredPower = -59;
```

Now we can use the startAdvertising provided by Bleacon to start advertising:

```
console.log('starting advertising ...');
Bleacon.startAdvertising(uuid, major, minor, measuredPower);
```

We are all set to run the *advertise.js* script now.

On a Mac:

```
$ node advertise.js
```

On Linux, sudo is needed:

```
$ sudo advertise.js
```

To detect the beacon, we need an application on our BLE-equipped smartphone. Go ahead and install the Locate Beacon app by Radius Networks on your smartphone:

- iOS (*http://bit.ly/Locate_Beacon*)

- Android (*http://bit.ly/1Sb0ugD*)

Once you open the app on iOS, you will be prompted for permission to access your location (as shown in Figure 10-1, left); make sure to select Allow. This prompt will not be present when using an Android device.

Now a prompt for sharing iBeacon data will appear (Figure 10-1, right). Select NO.

Figure 10-1 *Left: Location permissions screen of the Locate Beacon app; Right: Share iBeacon data screen of the Locate Beacon app*

The Main screen for the app is shown (Figure 10-2, left). Press the Locate iBeacons button to start the beacon-scanning process.

A screen showing that no beacons are in range will appear next (Figure 10-2, right), followed by a beacon being detected (Figure 10-3).

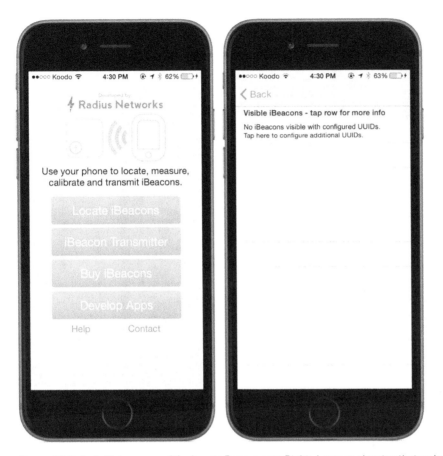

Figure 10-2 *Left: Main screen of the Locate Beacon app; Right: A screen showing that no beacons were found*

Figure 10-3 *Main screen of the Locate Beacon app*

 The UUID we selected, E2C56DB5-DFFB-48D2-B060-D0F5A71096E0, has been preconfigured in the Locate Beacon app. If a different UUID is used, the app would have to be configured to detect iBeacons using that UUID. This is only an issue on iOS, as the CoreLocation framework does not allow you to detect any iBeacon; Android does not have this issue.

Tap the row for a beacon to select it. You will see the iBeacon's RSSI and proximity (Figure 10-4, left). The proximity field can have one of the following values: immediate, near, and far. Walk further and closer to your iBeacon to see the proximity value change.

The measure-powered value for our beacon was arbitrarily selected. We can use the calibration feature of the Locate Beacon app to get a more accurate reading. Press the Calibrate button (Figure 10-4, right).

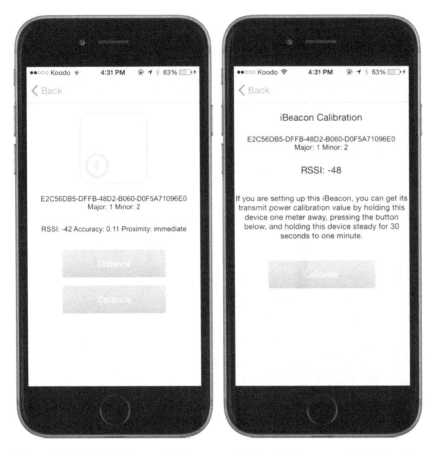

Figure 10-4 *Left: The iBeacon details screen of the Locate Beacon app; Right: The iBeacon Calibration screen of the Locate Beacon app*

Once you are about one meter away from your iBeacon, press the Calibrate button again to start the calibration progress, as shown in Figure 10-5 (left).

After 30 to 60 seconds, the app will recommend a measured-power value to use. In this particular session (Figure 10-5, right), -49 was recommended. Go ahead and update your *advertise.js* file with the recommended value. Then exit and rerun the script.

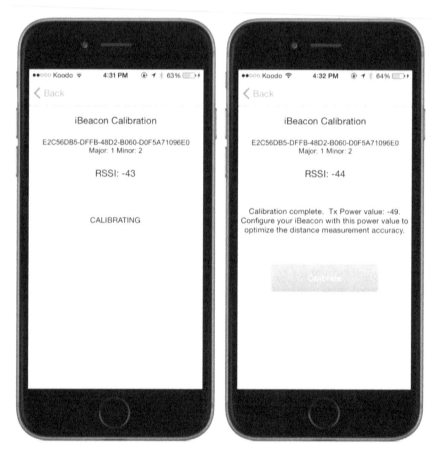

Figure 10-5 *Left: Screen showing that the iBeacon is calibrating; Right: Screen showing that the iBeacon was calibrated*

Creating a Mobile App that Uses iBeacons

Let's create a mobile app that detects the specific iBeacon we created earlier. We will use PhoneGap with an iBeacon plugin named `cordova-plugin-ibeacon`.

Source code and documentation for the `cordova-plugin-ibeacon` can be found on GitHub (*https://github.com/petermetz/cordova-plugin-ibeacon*).

There are two modes to detect iBeacons:

Ranging
> Used to detect iBeacons in range and periodically get their proximity (immediate, near, far). Ranging can only be used when the app is running in the foreground.

Region monitoring
> Used to detect if you are inside or outside an iBeacon region. Region monitoring can be used when the app is running in the foreground, as well as in background mode.

On iOS, both modes require the user to grant your app access to your location. Android apps only require permissions to access Bluetooth hardware.

The `cordova-plugin-ibeacon` plugin maps the modes to the following APIs for starting and stopping:

- `cordova.plugins.locationManager.startMonitoringForRegion(beaconRegion)`

- `cordova.plugins.locationManager.stopMonitoringForRegion(beaconRegion)`

- `cordova.plugins.locationManager.startRangingBeaconsInRegion(beaconRegion)`

- `cordova.plugins.locationManager.stopRangingBeaconsInRegion(beaconRegion)`

Both APIs require a `beaconRegion` parameter, which can be constructed from a `UUID`, `identifier`, optional `major` value, and optional `minor` value. If the major and/or minor is not provided, they are set to wildcard values so that any iBeacon with the UUID or UUID and major in range is reported.

```
var uuid = 'E2C56DB5-DFFB-48D2-B060-D0F5A71096E0';
var identifier = 'region';
var major = 1;
var minor = 2;

var beaconRegion = new cordova.plugins.locationManager.BeaconRegion(identifier, uuid, major, minor);
```

You also need to set up and assign a `delegate` to receive updates from the location manager. You have the option of overriding any of the default delegate methods.

A new location manager delegate can be created by using `new cordova.plugins.location Manager.Delegate()`. Then the default delegate methods for ranging and region monitoring can be over-ridden by setting the `didRangeBeaconsInRegion` and `didDetermineState ForRegion` properties. The `setDelegate` method of the location manager is then used to assign the delegate to the location manager.

```
var delegate = new cordova.plugins.locationManager.Delegate();

delegate.didRangeBeaconsInRegion = function(pluginResult) {
    // ...
};

delegate.didDetermineStateForRegion = function(pluginResult) {
    // ...
};

cordova.plugins.locationManager.setDelegate(delegate);
```

Let's create a region timer app to track how long we are in a beacon region. This could be used as a desk timer, for example. We only need to use the region-monitoring APIs for this app.

Start off by creating a new PhoneGap app, which we'll call `Region Timer`:

```
phonegap create regiontimer "com.makebluetooth.regiontimer" "Region Timer"
```

Then change directories to a newly created folder:

```
cd regiontimer
```

If you have an iOS device, add iOS as a platform using the following command:

```
phonegap platform add ios
```

To use with an Android device, use the following command:

```
phonegap platform add android
```

For Android, you will also need to update the *config.xml* file to update the `android-min SdkVersion` from 7 to 10 by changing the following line from:

```
<preference name="android-minSdkVersion" value="7" />
```

to:

```
<preference name="android-minSdkVersion" value="10" />
```

Now add the plug in to the PhoneGap app:

```
phonegap plugin add phonegap plugin add https://github.com/petermetz/cordova-plugin-
ibeacon.git#3.3.0
```

Open up *www/index.html* in a text editor and change the contents for the `<div class="app">` tag to:

```
<div id="totalTimeInRegion"></div>
```

We will use this `div` to display the total time spent in the region.

We are done with editing *index.html* for now. Now, to work on the JavaScript side, open *www/js/index.js* in a text editor.

Let's remove some of the code that we don't need anymore. Delete the `app.receivedE vent(deviceready);` line in the `onDeviceReady` function, as well as the entire `receivedE vent` function.

We need to create and set up a location manager as we did earlier. Create a new function named `setupLocationManager`:

```
setupLocationManager: function() {
}
```

The first thing we need to do is create a beacon region to monitor for the iBeacon UUID, major, and minor we created. We will use `region` as the identifier.

```
var uuid = 'E2C56DB5-DFFB-48D2-B060-D0F5A71096E0';
var identifier = 'region';
var major = 1;
var minor = 2;
```

```
var beaconRegion = new cordova.plugins.locationManager.BeaconRegion(identifier, uuid, ma
jor, minor);
```

Now create and assign a delegate to the Location Manager:

```
var delegate = new cordova.plugins.locationManager.Delegate();

cordova.plugins.locationManager.setDelegate(delegate);
```

On iOS 8 and above, we need to request authorization, including for background mode:

```
cordova.plugins.locationManager.requestAlwaysAuthorization();
```

Now we can start region monitoring:

```
cordova.plugins.locationManager.startMonitoringForRegion(beaconRegion);
```

The onDeviceReady also needs to be updated to call app.setupLocationManager to set up
the location manager:

```
// ...
onDeviceReady: function() {
    app.setupLocationManager();
},
// ...
```

We don't have anything to handle iBeacon region events, so we'll create a new function
called didDetermineStateForRegion. We can assign it to the delegate in the setupLoca
tionManager function:

```
setupLocationManager: function() {
    // ...
    var delegate = new cordova.plugins.locationManager.Delegate();
    delegate.didDetermineStateForRegion = app.didDetermineStateForRegion;


    // ...
}
```

Let's create the didDetermineStateForRegion function now. It will have one argument, plu
ginResult, which is a state property that contains the state of the region. When it is set to
'CLRegionStateInside', the device is in the beacon region; otherwise, the device is outside the region. We'll update the app.insideRegion value accordingly.

```
didDetermineStateForRegion: function(pluginResult) {
    var regionState = pluginResult.state;

    if (regionState === 'CLRegionStateInside') {
        app.insideRegion = true;
    } else {
        app.insideRegion = false;
    }
}
```

Now the `app.insideRegion` property reflects whether we are inside the iBeacon region. Let's create another function named `updateTotalTimeInRegion` that uses this property. It will first get the current time, and check if the user is in the region. When the user is in the region, it will update the total time in the region since the last update. The last update date will also be set to the current time.

```
updateTotalTimeInRegion: function() {
    var now = new Date();

    if (app.insideRegion) {
        var secondsSinceLastUpdate = (now.getTime() - app.lastUpdateDate.getTime()) /
1000.0;

        app.totalTimeInRegion += secondsSinceLastUpdate;
    }

    app.lastUpdateDate = now;
}
```

We also need to initialize the app properties in the `initialize` function:

```
// ...

initialize: function() {
    this.bindEvents();

    this.insideRegion = false;
    this.lastUpdateDate = null;
    this.totalTimeInRegion = 0;
},

// ...
```

Then we need to update `onDeviceReady` to set up a timer to call the new `app.updateTotal TimeInRegion` function every second (1,000 milliseconds). We can use `setInterval` for this.

```
// ...
onDeviceReady: function() {
    app.setupLocationManager();

    setInterval(app.updateTotalTimeInRegion, 1000);
},
// ...
```

This will ensure the total time in the region is updated every second, based on the value of `app.insideRegion`. If the user is not in the region, the `app.totalTimeInRegion` will remain the same. However, when the user is in the region, the value of `app.totalTimeInRegion` will be updated based on the time that has passed since the last update (usually one second).

We also need to update both `onDeviceReady` and `didDetermineStateForRegion` to call the `app.updateTotalTimeInRegion` when the app starts.

```
// ...
onDeviceReady: function() {
    // ...

    app.updateTotalTimeInRegion();
},
// ...
didDetermineStateForRegion: function(pluginResult) {
    // ...
    app.updateTotalTimeInRegion();
},
// ...
```

This will initialize things on app start and immediately update the time whenever the beacon region is entered or exited. Without this change, it would take about a second for the value to be updated.

Now app.totalTimeInRegion will be updated every time we enter and exit the beacon region. We need to display the total time in the app, so create a new function named displayTotalTimeInRegion.

The displayTotalTimeInRegion function will first split up totalTimeInRegion from seconds to hours, minutes, and seconds. Then create a string to display in the app, in the totalTimeInRegion element.

```
displayTotalTimeInRegion: function() {
    var hours = Math.floor(app.totalTimeInRegion / 3600);
    var minutes = Math.floor((app.totalTimeInRegion % 3600) / 60);
    var seconds = Math.floor(app.totalTimeInRegion % 60);

    var totalTimeInRegionText = 'Hours: ' + hours + ' ' +
                                'Minutes: ' + minutes + ' ' +
                                'Seconds: ' + seconds;

    document.getElementById('totalTimeInRegion').textContent = totalTimeInRegionText;
}
```

We are all set to run the initial version of our region timer app. Plug your iOS or Android device into your computer using a USB cable, and run the following command:

```
phonegap run --device
```

On iOS when the app launches, you will be prompted to allow location access (Figure 10-6). Press Allow to grant access; pressing Don't Allow will cause our app to not work.

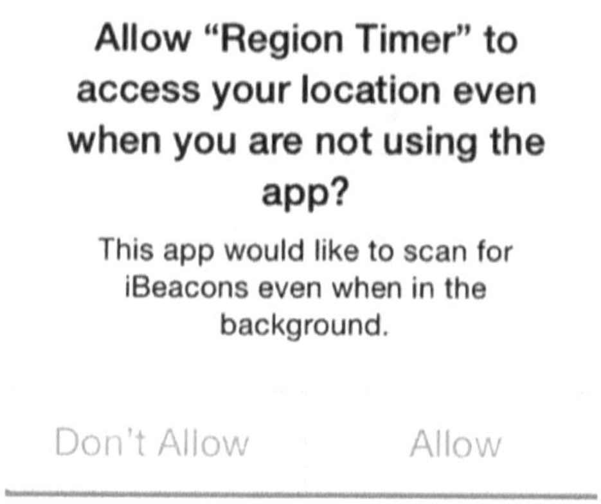

Allow "Region Timer" to access your location even when you are not using the app?

This app would like to scan for iBeacons even when in the background.

Don't Allow   Allow

Figure 10-6   *Location permissions screen of the Region Timer app*

The app will then go the the main screen, showing 0 time in the region (Figure 10-7).

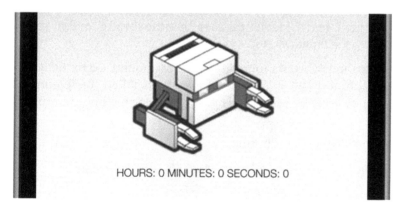

HOURS: 0 MINUTES: 0 SECONDS: 0

Figure 10-7   *The Region Timer app showing that our device is idle*

Once you start the Node.js iBeacon *advertise.js* script, the mobile app will start updating the time spent in the region (Figure 10-8).

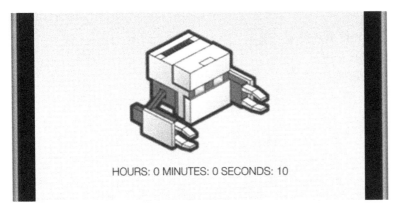

HOURS: 0 MINUTES: 0 SECONDS: 10

Figure 10-8   *The Region Timer app showing that our device is running*

After you stop *advertise.js*, it will take about 30 seconds for the location manager to report that you are no longer in the region. This is because of the built-in debounce logic.

You can restart *advertise.js* and experiment with the distance limits by walking further away from your beacon to get a better idea of the area that the iBeacon region covers. When you're far enough away, the timer will stop.

The app will also continue to work when it's running in the background. Go to the home-screen after launching the app, and walk in and out of the region. When you foreground the app, the time value will be updated accordingly. Keep in mind that it is only an esti-mate.

Let's add one final feature to the app, a reset button, to reset the time to zero.

Update the <div class="app"> in *www/index.html* so it looks like the following:

```
<div class="app">
    <div id="totalTimeInRegion"></div>
    <button id="reset">Reset</button> ❶
</div>
```

❶   New reset button

Then update the onDeviceReady function in *www/js/index.js* and add a click event listener for the reset button. We'll make a new function named onResetClick to handle the event. The onResetClick will reset app.totalTimeInRegion to 0, and then call app.updateTotalTi meInRegion.

```
// ...
onDeviceReady: function() {
    // ...
    document.getElementById('reset').addEventListener('click', app.onResetClick, false);
},
// ...
onResetClick: function() {
    app.totalTimeInRegion = 0;
```

```
        app.updateTotalTimeInRegion();
    }
```

When you run the updated app, a reset button will be displayed under the timer value (Figure 10-9).

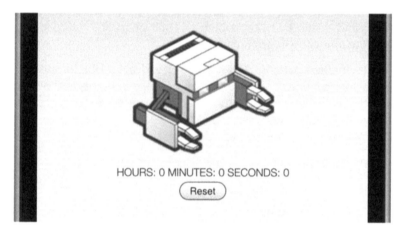

HOURS: 0 MINUTES: 0 SECONDS: 0

Reset

Figure 10-9 *The main screen of the Region Timer app with a reset button*

The time in the region will be reset back to 0 when the button is tapped.

Eddystone Beacons and the Physical Web

Eddystone is cross-platform and open-source beacon specification for the BLE proximity beacon and is distributed under the Apache v2.0 license. On Android, it is built into the Google Play Services Nearby API. On iOS, Eddystone beacon support is added using a library.

Eddystone supports multiple message types, including:

Eddystone-UID
> Broadcasts a 16-byte beacon ID, which contains a 10-byte namespace ID and a 16-byte instance ID. The namespace ID is intended to be used to group a set of beacons, and the instance ID identifies individual devices in the group. This is a similar concept to iBeacon, which we covered earlier in the chapter.

Eddystone-URL
> Broadcasts a compressed URL, which is needed because of the limited space available in an advertisement packet. Once a central discovers a beacon, it has to uncompress the URL before it is of any use.

Eddystone-TLM
> Broadcasts telemetry data about the beacon. This includes battery voltage and device temperature.

The full Eddystone specification can be found on GitHub (*https://github.com/google/eddystone*).

The Physical Web's (*http://google.github.io/physical-web/*) project slogan is "Walk up and use anything." The goal of the project is to enable frictionless discovery of web content that is related to your surroundings. You should be able to walk up to any smart device, such as a vending machine or toy, and not have to download an app to use it. It is based on Eddystone-URL's beacons and existing web technologies. Each smart device broadcasts a URL, which provides the UI to interact with that device.

For example, a Physical-Web–enabled parking meter would broadcast a URL for a cloud-based web application. Your smartphone would scan for all beacons in range, and display a list of them, sorting by the ones in closest proximity. When the parking meter is in range and you decide to interact with it, you would visit the URL it broadcasts in a mobile web browser, and pay for your parking—all without downloading a specific mobile app!

You will still need an app installed on your smartphone to use the Physical Web.

What Data Does an Eddystone Beacon Advertise?

Unlike iBeacon, Eddystone beacons use the service data type to broadcast information during advertisement.

The specification requires the following to be broadcasted:

- GAP flags with a value of `0x06`
- A service UUID of `0xFEAA`
- Service data for the `0xFEAA` UUID

The service data has a different format for each sub-beacon type.

Eddystone-UID

For Eddystone-UID beacons, service data contains the following 20-byte structure:

- The first byte is the frame type, which has a value of `0x00`
- The second byte is the calibrated TX power at 0 meters
- The next 10 bytes contain the Namespace ID
- The next 6 bytes contain the Instance ID
- The final 2 bytes must be `0x00`, and are reserved for future use

Here's an example of service data for an Eddystone-UID beacon, followed by a description of each of its parts:

```
00 EC 010203040506070809010 AABBCCDDEEFF 0000
```

- The first bytes are the frame type of 0x00 for the UID

- The next 2 bytes are the calibrated TX power at 0 meters over -20 dBm (0xEC)
- Next is the namespace ID of 01020304050607080910
- The fourth set of bytes contains the instance ID of AABBCCDDEEFF
- Reserved bytes of 0000

Eddystone-URL

For Eddystone-URL beacons, service data contains the following byte structure, which can be up to 20 bytes in length:

- The first byte is the frame type, which has a value of 0x10
- The second byte is the calibrated TX power at 0 meters
- The third byte contains the encoded URL scheme
- The remaining bytes contain the encoded URL

Values for the encoded URL scheme are:

Value	Expansion
0x00	http://www.
0x01	https://www.
0x02	http://
0x03	https://

The encoded URL value can uses the following codes to reduce its size:

Value	Expansion
0x00	.com/
0x01	.org/
0x02	.edu/
0x03	.net/
0x04	.info/
0x05	.biz/
0x06	.gov/

Value	Expansion
0x07	.com
0x08	.org
0x09	.edu
0x0a	.net
0x0b	.info
0x0c	.biz
0x0d	.gov

Here's an example of service data for an Eddystone-URL beacon, followed by a description of each of its parts:

```
10 EE 0061626307
```

- The first 2 bytes are the frame type of 0x10
- The next 2 bytes are the calibrated TX power at 0 meters over -18 dBm (0xEE)
- The third set of bytes contains the encoded URL scheme of *http://www* . (0x00)
- abc encoded as ASCII (616363)
- .com extension (0x07 from the preceding table)

This creates an Eddystone-URL beacon that broadcasts *http://www.abc.com*.

Building and Detecting Your Own Beacon

We'll be focusing on Eddystone-URL beacons for the remainder of the chapter.

Let's create an Eddystone-URL beacon using Node.js and use a smartphone to detect it.

We'll use the node-eddystone-beacon module using Node.js. It uses bleno for the BLE layer.

The source code for node-eddystone-beacon can be found on the following GitHub repo (*https://github.com/don/node-eddystone-beacon*).

First, make a new directory for the project, and change the current directory to it:

```
$ mkdir make-bluetooth-physical-web
$ cd make-bluetooth-physical-web
```

Next, install the eddystone-beacon module from npm:

```
$ npm install eddystone-beacon
```

This command will create a node_modules folder in the current directory and pull the eddy stone-beacon module along with its dependencies, including bleno, down from npm.

Now create a new file called *advertise.js* and open it using your favorite text editor.

At the start of the file, we need to require the eddystone-beacon module:

```
var EddystoneBeacon = require('eddystone-beacon');
```

Next, let's create a variable for the URL we wish to broadcast using the Eddystone-URL format:

```
var url = 'http://example.com';
```

Now we can use the advertiseUrl API provided by EddystoneBeacon to start advertising the URL:

```
EddystoneBeacon.advertiseUrl(url);
```

We are all set to run the *advertise.js* script now.

On a Mac:

```
$ node advertise.js
```

On Linux, sudo is needed:

```
$ sudo node advertise.js
```

To detect the beacon, we need an application on our BLE-equipped smartphone. Go ahead and install the Physical Web app on your smartphone:

- Physical Web for iOS (*http://apple.co/1Sb3c5H*)
- Physical Web for Android (*http://bit.ly/1Sb3cCR*)

When you first open the Physical Web application, a brief welcome screen is shown (Figure 10-10, left). Tap the Proceed button to continue.

When no Eddystone-URL beacons are in range, Figure 10-10 (right) is displayed.

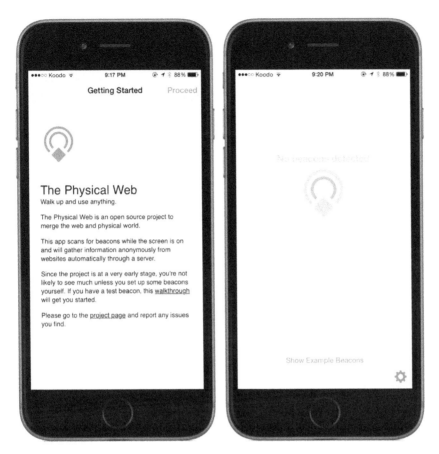

Figure 10-10 *Left: Welcome screen for Physical Web app; Right: Physical Web app screen showing no beacons in range*

Since we are running *advertise.js*, at least one beacon will be in range. If multiple beacons are in range, they will all be displayed in the list and sorted by signal strength (Figure 10-11, left).

Go ahead and tap the Example Domain beacon item. This will open a web browser to *http://example.com* (Figure 10-11, right), the URL that our Eddystone-URL beacon is configured to advertise.

Unlike iBeacon, which we tried out before, you do not need a specific app to utilize Eddystone-URL beacons; the Physical Web app scans for all Eddystone-URL beacons in range. Then it uses a mobile web browser to provide the UI to interact with the device, once you select a URL you are interested in. Our simple example doesn't provide any interaction as the example.com website is very static. However, we were able to create a beacon that advertises the URL physically, allowing anyone with an app like Physical Web to detect the beacon when they are in range.

Figure 10-11 *Left: Physical Web app showing beacon in range; Right: Physical Web app displaying http://example.com*

Conclusion

In this chapter, we created two beacons using Node.js, and used a BLE-enabled smartphone with the appropriate app installed to detect the beacon we created. We used the iBeacon and Eddystone-URL beacon format, which both rely on the GAP portion of the BLE standard.

We also created a custom region timer mobile app using PhoneGap and an iBeacon plugin that measures the time spent in an iBeacon region.

Drones | 11

In this chapter, we will control a Parrot Rolling Spider (*http://www.parrot.com/ca/products/rolling-spider/*) drone using a computer.

The Rolling Spider is a BLE-controlled ultra-compact drone, which is part of Parrot's Mini Drone product line. It is lifted and propelled by four rotors, which classifies it as a quadcopter. See Figure 11-1 for a photo of a Rolling Spider with wheels attached.

The NodeCopter (*http://www.nodecopter.com*) community initially started off using the Parrot AR.Drone 2.0 (*http://ardrone2.parrot.com*) for hackathon events around the world. Participants would create Node.js applications to control the drone using node-ar-drone (*https://github.com/felixge/node-ar-drone*). Unlike the Rolling Spider, the AR.Drone 2.0 is a much larger drone, that is controlled over WiFi.

We will be using the node-rolling-spider (*https://github.com/voodootikigod/node-rolling-spider*) module to control the Rolling Spider. It offers a similar API as node-ar-drone, but supports the Rolling Spider.

Figure 11-1 *Parrot Rolling Spider with wheels on*

What You'll Need

- Parrot Rolling Spider (*http://www.parrot.com/ca/products/rolling-spider/*), available at the MakerShed store (*http://www.makershed.com/products/parrot-rolling-spider*)

- One of the following:

  — Bluetooth-4.0–enabled Mac

  — Linux PC (includes Raspberry Pi or Beagle Bone Black) with a Bluetooth 4.0 adapter, such as the Bluetooth 4.0 USB Module (*https://www.adafruit.com/products/1327*) available on Adafruit

- Bluetooth-4.0–enabled smartphone or tablet running iOS or Android (for testing)

Node.js and noble will also need to be set up; see "Installing Node.js" in Chapter 2 for instructions.

Testing Out the Drone

Follow the instructions supplied by Parrot to assemble your Rolling Spider drone and charge the battery. We highly recommend using the drone with the wheels on!

Then test it out by installing and using the Parrot FreeFlight 3 app from the App Store or Google Play Store on your BLE-equipped smartphone.

Controlling the Rolling Spider with Node.js

We'll use the `node-rolling-spider` module to control the Rolling Spider drone from Node.js. It uses `noble` for the BLE communication layer. The module was started by Jack Watson-Hamblin (@FluffyJack (*https://github.com/FluffyJack*)). Chris Williams (@voodootiki-god (*https://github.com/voodootikigod*)) is now the active maintainer of the project.

The source code for `node-rolling-spider` can be found at GitHub repo (*https://github.com/voodootikigod/node-rolling-spider*).

Setting Up the Project

Now that we have played with the smartphone app, let's move on to controlling the drone with Node.js.

First, make a new directory for the project, and change the current directory to it:

```
$ mkdir make-bluetooth-rolling-spider
$ cd make-bluetooth-rolling-spider
```

Next, install the `rolling-spider` module from npm:

```
$ npm install rolling-spider
```

This command will create a node_modules folder in the current directory and pull the `rolling-spider` module, along with its dependencies including `noble`, down from npm.

Discovering the Drone

`node-rolling-spider` includes an example *discovery.js* file we can use to verify that everything is set up correctly.

On a Mac:

```
$ node node_modules/rolling-spider/eg/discover.js
```

On Linux, sudo is needed:

```
$ sudo node node_modules/rolling-spider/eg/discover.js
```

If your system is set up correctly, something similar to the following will be output to the command line.

On a Mac:

```
1: RS_W169095 (39d6c56f47824599b3a1b89274d99630), RSSI -50
```

On Linux:

```
1: RS_W169095 (e014b1ae3d4e), RSSI -93
```

The script searches for Rolling Spider drones in range and outputs the info the drone is advertising:

local name
> Can be set via the smartphone app. The default local name indicates the color of the drone; see the first letter after the RS_ section: W(hite), R(ed), or B(lue)

id (in brackets)
> ID of the drone, as reported by noble

RSSI
> Signal strength of the drone; a larger number (closer to zero) indicates the drone is closer than a smaller number (more negative)

Getting Started: Basic Takeoff and Landing

Create a new file named *takeoff-land.js* using your favorite text editor in the project directory you set up earlier.

The first thing you need to do is require the rolling-spider module:

```
var RollingSpider = require('rolling-spider');
```

Next, create an new instance of RollingSpider:

```
var rs = new RollingSpider();
```

 The default behavior of the module is to connect to the first Rolling Spider discovered. If more than one Rolling Spider is in range, you can pass the id of the drone in; for example:

```
var rs = new RollingSpider('e014b1ae3d4e');
```

The id can be determined using the discover.js example script we used earlier.

Now we need to get our RollingSpider instance, rs, to connect. Let's create a new function named connect and call the connect method on our Rolling Spider instance, passing in a callback function named connectCallback, which we will create next. We can then call connect to start the connection sequence.

```
function connect() {
  console.log('connecting ...');
```

```
    rs.connect(connectCallback);
  }

  connect();
```

Now to implement the connectCallback function, it has an error parameter that indicates if there was an error while trying to connect to the Rolling Spider. When an error occurs, we will just exit. If the connection was successful, we then need to call the setup method of our Rolling Spider instance. Again, we'll pass a callback named setupCallback, which we need to create next.

```
  function connectCallback(error) {
    if (error) {
      console.log('error connecting: ' + error);
      process.exit(-1);
    }

    console.log('connected, setting up ...');
    rs.setup(setupCallback);
  }
```

In the setupCallback function, we need to call the flatTrim and startPing methods. The flatTrim API calibrates the drone's sensors so it knows what the level/flat value is before takeoff. It should always be called before taking off.

```
  function setupCallback() {
    console.log('set up, flat trimming and starting ping ...');

    rs.flatTrim();
    rs.startPing();
    rs.flatTrim();
  }
```

We are all set to take off, but we must add a delay for the Rolling Spider to proccess the flatTrim and startPing commands. We use setTimeout to delay calling the new takeOff function by 1000ms, or 1 second.

```
  function setupCallback() {
    // ...

    setTimeout(takeOff, 1000);
  }
```

In the takeOff function, call the takeOff method on our Rolling Spider instance, passing in the takeOffCallback function. For now, the takeOffCallback function will only log to console.

```
  function takeOff() {
    console.log('taking off ...');
    rs.takeOff(takeOffCallback);
  }

  function takeOffCallback() {
```

```
    console.log('taken off');
  }
```

We have almost finished a basic takeoff and landing sequence. After the Rolling Spider has taken off, we need to land it after a delay. Let's update the takeOffCallback by adding code for landing.

We use setTimeout again to call the land function after a two-second delay.

```
function takeOffCallback() {
  // ...
  setTimeout(land, 2000);
}
```

In the land function, call the land method on the Rolling Spider instance, with the land Callback as a parameter. The landCallback will be executed when the Rolling Spider has landed, and we can exit when this occurs.

```
function land() {
  console.log('landing ...');
  rs.land(landCallback);
}

function landCallback() {
  console.log('landed')
  process.exit(0);
}
```

Now we are ready to try out our basic takeoff and land script:

On a Mac:

```
$ node takeoff-land.js
```

On Linux, sudo is needed:

```
$ sudo node takeoff-land.js
```

While the Rolling Spider takes off and lands shortly thereafter, the script will output the following:

```
connecting ...
connected, setting up ...
set up, flat trimming and starting ping ...
taking off ...
taken off
landing ...
landed
```

Keyboard Control

Now that we are familiar with the node-rolling-spider APIs, let's create a Node.js applica-tion to control the Rolling Spider using keyboard input.

The Node.js keypress module allows you to easily listen to keyboard events from the `stdin` input stream. Nathan Rajlich (@TooTallNate (*https://github.com/TooTallNate*)) wrote the key press module.

Source code for keypress can be found here on GitHub (*https://github.com/TooTallNate/keypress*).

Let's install the module using npm:

```
$ npm install keypress
```

Now create a new file named *keypress-test.js*.

The first step is to require the keypress module:

```
var keypress = require('keypress');
```

Then we can set up the keypress module on the `stdin` stream:

```
keypress(process.stdin);
```

We also need to enable raw mode on the `stdin` stream. Without raw mode, the keyboard input will be buffered so you would not get the keypress event immediately.

```
process.stdin.setRawMode(true);
```

The last step is to attach an event handler for the keypress event. First, we need to check if the key argument is provided; if it isn't, we can stop processing the event and return. The key argument is not present for some keys, like number keys. Then we can check if Ctrl-C was entered. This is usually handled for you, but since we've set up the keypress module, we need to handle it ourselves. Otherwise, we will print the keypress event key to the console.

```
process.stdin.on('keypress', function (ch, key) {
  if (!key) {
    return;
  }

  if (key.ctrl && key.name === 'c') {
    process.exit();
  }

  console.log('got "keypress"', key);
});
```

Now we can run the application:

```
$ node keypress-test.js
```

When you press the up arrow key on your keyboard, the test application will output the following:

```
got "keypress" { name: 'up',
  ctrl: false,
```

```
meta: false,
shift: false,
sequence: '\u001b[A',
code: '[A' }
```

The name property of the key has the value of up; we can ignore the other properties.

You can try other key presses and see the key event output. Press the Ctrl-C at the same time to exit.

Let's update the keypress event handler to only log the name of the key that was pressed:

```
process.stdin.on('keypress', function (ch, key) {
  // ...

  console.log('got "keypress"', key.name);
});
```

When we run the application again:

```
$ node keypress-test.js
```

You will see a simplified output of key events, like the following for the up arrow key:

```
got "keypress" up
```

If you hold a key down many events are received, this will cause issues when you map keypress events to Rolling Spider actions. Too many commands will be sent to the drone when a key is held down.

Let's add a debouncing mechanism to the keypress event handler to throttle out event handling. We start by adding a global variable to track the state at the top of the file:

```
var active = false;
```

Then we check the value in the keypress event handler. If active is true, we abort processing the event. Otherwise, we can set active to true.

```
process.stdin.on('keypress', function (ch, key) {
  // ...

  if (active) {
    return;
  }
  active = true;

  // ...
});
```

Now we need to set the value of active to false after a delay. We can use the setTimeout function to accomplish this, with a delay of 100 milliseconds.

Create a new function named clearActive that sets the active variable value to false. At the end of the keypress event handler, use setTimeout to call clearActive after 100 ms.

```
function clearActive() {
  active = false;
}

process.stdin.on('keypress', function (ch, key) {
  // ...

  setTimeout(clearActive, 100);
});
```

Now when we run the test application again, and hold a key down, there will be a 100-ms delay between events when they are logged to the console.

```
$ node keypress-test.js
```

Putting it all together

Now that we have familarized ourselves with the `node-rolling-spider` and `keypress` modules, we can combine them.

The code from earlier can be reused with a few changes. We don't want to automatically have the Rolling Spider take off and land.

Create a new file named *keyboard.js* with the following starter code:

```
var RollingSpider = require('rolling-spider');

var rs = new RollingSpider();

function connect() {
  console.log('connecting ...');
  rs.connect(connectCallback);
}

function connectCallback(error) {
  if (error) {
    console.log('error connecting: ' + error);
    process.exit(-1);
  }

  console.log('connected, setting up ...');
  rs.setup(setupCallback);
}

function setupCallback() {
  console.log('set up, flat trimming and starting ping ...');

  rs.flatTrim();
  rs.startPing();
  rs.flatTrim();

  setTimeout(ready, 1000);
}

function ready() {
```

```
  // ...
}
```

```
connect();
```

When the Rolling Spider is ready to receive commands, we'll call a new function named ready. Previously, we had code to take off and land. Now let's add the keypress functionality. At the start of the file, add the following:

```
var keypress = require('keypress');
var RollingSpider = require('rolling-spider');

keypress(process.stdin);

var active = true;
```

This is similar to what we used earlier when trying out the keypress module. We'll also set up a variable named active with an initial value of true to ignore keypress events until the Rolling Spider is ready to receive commands.

Now we can add the remaining keypress handling at the end of the file. Again, this is similiar to what we tried out before.

```
  // ...

process.stdin.setRawMode(true);

function clearActive() {
  active = false;
}

process.stdin.on('keypress', function (ch, key) {
  if (!key) {
    return;
  }

  if (key.ctrl && key.name === 'c') {
    process.exit();
  }

  if (active) {
    return;
  }
  active = true;

  setTimeout(clearActive, 100);
});

connect();
```

Now we will fill in the ready function we created earlier. We'll print a message to the console to let the user know things are ready. Then we will clear the active flag to enable key press event handling.

```
function ready() {
  console.log('ready to receive commands!');
  console.log();

  clearActive();
}
```

Now let's add some code to map keypress events to Rolling Spider commands. We'll map the t key to takeOff and l to land.

```
process.stdin.on('keypress', function (ch, key) {
  // ...

  if (key.name === 't') {
    console.log('taking off');
    rs.takeOff();
  } else if (key.name === 'l') {
    console.log('landing');
    rs.land();
  }

  setTimeout(clearActive, 100);
});
```

Now we are ready to try taking off and landing using keyboard controls.

On OS X:

```
$ node keyboard.js
```

On Linux, sudo is needed:

```
$ sudo node keyboard.js
```

Wait for the following output:

```
connecting ...
connected, setting up ...
set up, flat trimming and starting ping ...
ready to receive commands!
```

Now press the t key to make the Rolling Spider take off. Once the Rolling Spider is in the air, wait a few seconds, and then press l to initiate landing. Then use Ctrl-C to exit the application.

Let's map the keyboard arrow keys to Rolling Spider commands. The APIs we'll use for this take an options parameter, which can contain the following properties:

speed
> The speed to use for the drive or rotation; a number between 0 and 100 inclusively

steps
> The step (time) to use for the drive or rotation; a number between 0 and 100 inclusively

We'll use a step size of 2 and a default speed for the new commands so we need to create an options variable that will be passed into the action functions. The up and down keys will be mapped to up and down, respectively. The left and right keys will make the Rolling Spider turn left and right.

```
process.stdin.on('keypress', function (ch, key) {
  // ...

  var options = {
    steps: 2
  };

  if (key.name === 't') {
    console.log('taking off');
    rs.takeOff();
  } else if (key.name === 'l') {
    console.log('landing');
    rs.land();
  } else if (key.name === 'up') {
    console.log('up');
    rs.up(options);
  } else if (key.name ==='down') {
    console.log('down');
    rs.down(options);
  } else if (key.name === 'left') {
    console.log('turn left');
    rs.turnLeft(options);
  } else if (key.name ==='right') {
    console.log('turn right');
    rs.turnRight(options);
  }

  // ...
});
```

Now let's add some commands for moving forward, backward, left, and right. The w key will be used for forward, s for backward, a for left, and d for right.

```
process.stdin.on('keypress', function (ch, key) {
  // ...

  // ...
  } else if (key.name === 'w') {
    console.log('forward');
    rs.forward(options);
  } else if (key.name ==='s') {
    console.log('backward');
    rs.backward(options);
  } else if (key.name === 'a') {
    console.log('left');
    rs.left(options);
  } else if (key.name ==='d') {
    console.log('right');
```

```
    rs.right(options);
  }

  // ...
});
```

See Table 11-1 for a summary of the key mappings we implemented.

Table 11-1 *Summary of keyboard commands*

Key	Action
t	take off
l	land
up arrow	up
down arrow	down
left arrow	turn left
right arrow	turn right
w	forward
s	backwards
a	left
d	right

Time to try out the commands!

Like before, wait for the "ready to receive commands!" message. Then press t to take off, and once you're in the air, use the keys you mapped to actions to move the Rolling Spider around. Press l to land the drone.

Additional Rolling Spider commands, such as tilt left/right, can also be mapped to key press events, but we will stop here.

Conclusion

In this chapter, we created a Node.js application to control a Rolling Spider using keyboard input. This was done using the node-rolling-spider and keypress modules. The Node.js code can be expanded to map other keypress events to commands, such as flips (front, back) and tilts.

Going Further | 12

This book can only serve as an introduction. It has walked you through getting started with building your own hardware and writing Bluetooth services. But there is a lot more to learn.

The Arduino

The Arduino is an amazingly flexible platform, and like several other topics in this book we really haven't gone into huge amounts of detail about its capabilities. If you want to learn more, you should probably take a look at the *Arduino Cookbook, 2e* by Micahel Margolils (O'Reilly), or the *Arduino: Up and Running* video course with Brian Jepson (O'Reilly).

Hardware Suggestions

Throughout the book we've made use of the Adafruit Bluefruit LE (*http://www.adafruit.com/products/1697*) board based around the Nordic Semiconductor nRF8001 chipset. We used that board because the nRF8001 is among the simplest chipsets to deal with, and it has good library support (*https://github.com/sandeepmistry/arduino-BLEPeripheral*) (see Figure 2-5) to build custom Bluetooth LE services with the Arduino.

But it's certainly not the only Bluetooth LE hardware we'd recommend. In Chapter 4 we also made use of the RedBearLab BLE Shield (*http://redbearlab.com/bleshield/*) as an alternative, and we recommend that and other RedBearLab Bluetooth LE boards.

Figure 12-1 shows the RedBearLab nRF51822 (left), the Blend (middle-left), and the Blend Micro (middle-right), along with the BLE Nano and adaptor board (right). All of these boards are based around the nRF8001 or nRF51822 chipsets and are compatible with the BLE Peripheral Library (see "Installing the BLE Peripheral Library").

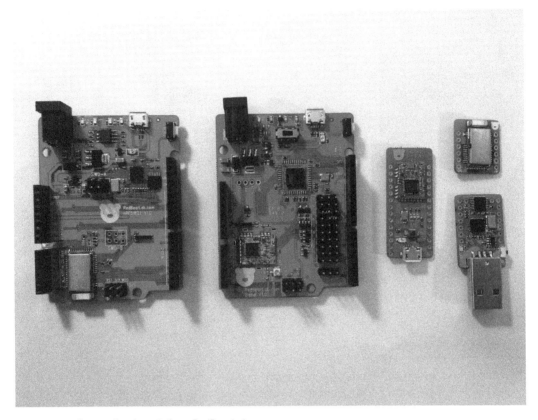

Figure 12-1 *Some other boards from RedBearLab*

Another board that's compatible with the BLE Peripheral Library is the RFduino (*http://www.rfduino.com/*), whose small form factor and relatively low cost make it useful for a lot of projects.

Unfortunately it's hard to recommend individual boards as they come and go fairly frequently. Many of the boards we would have recommended just a few months ago are no longer in production. That's one of the reasons we used the Arduino Uno and a board based around the nRF8001 chipset throughout the book. You can easily replace the Arduino Uno with an Arduino Leonardo or a dozen other compatible boards, and there are a number of boards (like those from RedBearLab) that use the nRF8001 chipset and can be substituted for Adafruit's Bluefruit breakout board without having to make changes to the code.

For an updated board list, check the compatible hardware section of the arduino-BLEPeripheral project page (http://bit.ly/1Sb7KJg).

Beyond that, there is a lot of other hardware available, much of which isn't—at least not yet—compatible with the BLE Peripheral library. One good example is the boards from Punch Through Design.

The LightBlue Bean and Bean+ boards (see Figure 12-2) from Punch Through Design allow you to load sketches onto a board over-the-air using Bluetooth LE. They also have have the ability to write, and then upload, code directly from your phone over Bluetooth LE.

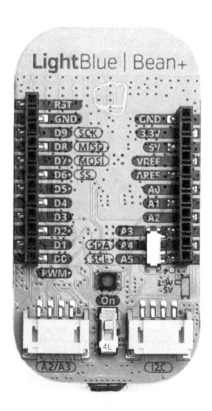

Figure 12-2 *The LightBlue Bean (left), and Bean+ (right)*

Unfortunately, the Beans are not compatible with the BLE Peripheral Library, as Punch Through's LBM313 module is based around the TI CC2540 chip. They also use a UART model instead of allowing you to implement real custom services, although the Bean does have five "scratch" characteristics that can be used to read and write (though not yet notify) arbitrary data.

Despite this, the hardware is solid and their Bean library (*http://bit.ly/1Sb7ZUC*) for the Arduino is excellent, and due to the architecture of the board, that puts the LBM313 module in the middle. The Bean library includes a `Bean.sleep()` function you can use to put the power-hungry ATmega328p—the Arduino—to sleep (*http://bit.ly/1Sb80rG*).

The lack of custom services is a big hit against the Beans, but right now they are the only Arduino-compatible boards that support writing, compiling, and then uploading a sketch directly from any sort of mobile platform.

Further Reading

This book was aimed squarely at introducing you to the practical side of Bluetooth LE through building a few interesting projects. As a result, there is a lot we haven't talked about.

If you're looking for a good introduction to the standard, we recommend *Getting Started with Bluetooth Low Energy* by Kevin Townsend, Carles Cufí, Akiba, and Robert Davidson (O'Reilly). Despite the title, the book takes a far more academic approach to Bluetooth LE and goes into a lot more detail about the protocol design than we do.

However, if you're interested in taking a deep dive into the Bluetooth LE standard without sitting down to read the thousands of pages of the core specification, then we recommend *Bluetooth Low Energy* by Robin Heydon (Prentice-Hall). The book has served as the bible for the protocol for at least one of this book's authors, and if you really want to know how Bluetooth LE works, then this book is for you.

HID Over GATT Pairing

This appendix walks through how to pair and unpair your HID over a GATT peripheral with various central devices.

iOS

Pairing

1. Open the Settings app (Figure A-1, left).

2. Go to Bluetooth (Figure A-1, right).

3. Ensure that Bluetooth is on, and touch the Arduino device. Select Pair when prompted (Figure A-2, left).

4. The device is now connected (Figure A-2, right).

Figure A-1 Left: iOS Settings; Right: iOS Bluetooth Settings

Figure A-2 *Left: iOS Bluetooth Pairing Request; Right: iOS Bluetooth Settings screen showing that it is connected to the Arduino*

Unpairing

1. Go back into Settings → Bluetooth.

2. Touch the i in the circle on the right side of Arduino.

3. Touch "Forget This Device" (Figure A-3, left).

4. Confirm by touching "Forget Device" (Figure A-3, right).

Figure A-3 Left: Unpairing a device on iOS; Right: Confirming that you want to unpair a device on iOS

Android

Pairing

1. Open the Settings app (Figure A-4, left).

2. Go to Bluetooth (Figure A-4, right).

3. Ensure that Bluetooth is on, and touch the HID Volume device (Figure A-5, left).

4. The device is now connected (Figure A-5, right).

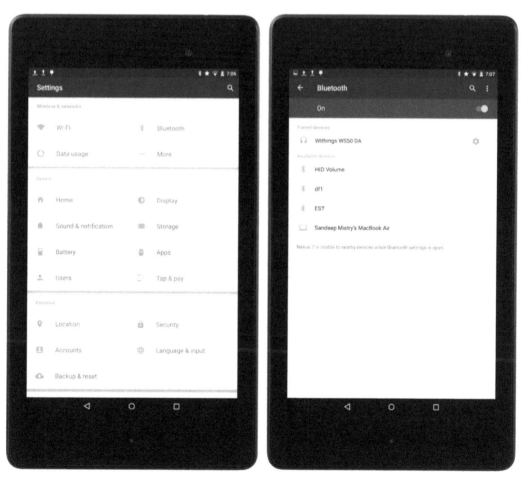

Figure A-4 *Left: Android Settings; Right: Android Bluetooth Settings*

Figure A-5 *Android Bluetooth device-pairing screen*

Unpairing

1. Go back into Settings → Bluetooth.

2. Touch the gear icon on the right side of HID Volume.

3. Touch "Forget" (Figure A-6).

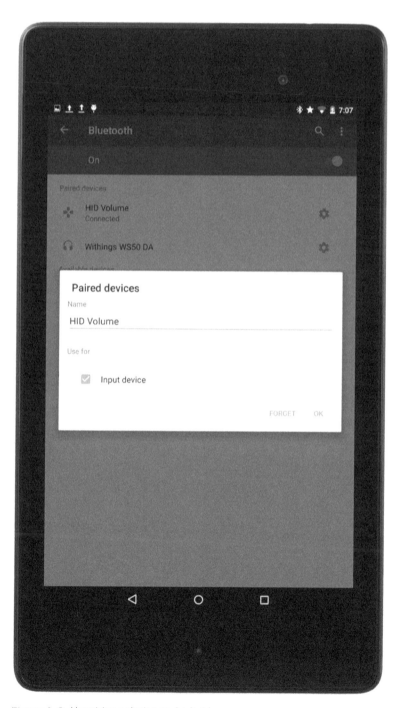

Figure A-6 *Unpairing a device on Android*

OS X

Pairing

1. Open the System Preferences app (Figure A-7).

2. Go to Bluetooth (Figure A-8).

3. Ensure Bluetooth is on, and click the pair button beside the HID Volume device.

4. The device is connected (Figure A-9).

Figure A-7 *OS X System Preferences*

Figure A-8 OS X Bluetooth Settings

Figure A-9 OS X screen showing paired devices

Unpairing

1. Go back to System Preferences → Bluetooth.

2. Click on the Arduino device (Figure A-10).

3. Click the X on the right side.

4. Click the Remove button to confirm unpairing (Figure A-11).

Figure A-10 *Click the X next to the device you want to unpair*

Figure A-11 *Confirming that you want to unpair the device*

Index

U

Uint8Array, 132
Universally Unique Identifiers (UUIDs), 4-5
unlock function, 73
UnlockCharacteristic function, 86, 88
unlockCharacteristicWritten function, 59
updateLights function, 124, 142

V

volume knob, 170-180
    Arduino library setup, 173
    hardware and wiring, 171-171
    testing rotary encoder, 173-174

W

weather station, 93-116

connecting with iOS, 101
hardware and wiring, 94-95
libraries, 96
PhoneGap app for, 104-112
    create the project, 104
    HTML in, 105-106
    JavaScript in, 107-112
    running, 115
programming, 96-99
Serial Monitor, 101
Windows
    Arduino IDE installation, 13
    BLE support, 7
write command property, 4
write request property, 4

About the Authors

Alasdair Allan is a scientist, author, hacker, and tinkerer. He is the author of a number of books (*http://amzn.to/1Yfjrll*), and sometimes also stands in front of cameras. You can often find him at conferences talking about interesting things, or deploying sensors (*http://data sensinglab.com*) to measure them. A couple of years ago, during Google I/O, he rolled out a mesh network (*http://bit.ly/1lGVVQm*) of 400 sensor motes covering the entire Moscone West. A few years before that he caused a privacy scandal (*http://oreil.ly/1Ll5grJ*) by uncovering the fact that your iPhone was recording your location all the time. This caused several class action lawsuits and a U.S. Senate hearing. He is a contributing editor for *Make: Magazine* (*http://makezine.com*), writing about electronics—especially wireless devices and distributed sensor networks, mobile computing, and the "Internet of Things"—and a former astrophysicist, where he contributed to the detection of what was—at the time—the most distant object yet discovered (*http://arxiv.org/abs/0906.1577*).

Don Coleman is a lifelong engineer who has come full circle, from mechanical to software and now to hardware, bridging the gap between all disciplines. Don is a seasoned Phone-Gap developer who has embraced it since its inception, and has spoken across the country about the benefits and advantages of using PhoneGap. As the Director of Consulting for Chariot Solutions, a software consulting company near Philadelphia, PA, he works with teams and clients to reinvent their existing technology and lay the groundwork for the future.

Sandeep Mistry is a professional software engineer who enjoys tinkering with the Internet of Things and Bluetooth Low Energy (BLE) devices. Sandeep has created/authored numerous open source BLE libraries, including noble and bleno for Node.js, and BLEPeripheral for Arduino.

Colophon

The cover photo is from Ken Rimple. The cover fonts are URW Typewriter and Guardian Sans. The text font is Adobe Minion Pro; the heading font is Adobe Myriad Condensed; and the code font is Dalton Maag's Ubuntu Mono.

CPSIA information can be obtained
at www.ICGtesting.com
Printed in the USA
LVHW07s0239230518
578092LV00003B/14/P

9 781457 187094